设计智能
高级计算性建筑生形研究
DESIGN INTELLIGENCE
ADVANCED COMPUTATIONAL RESEARCH

学生建筑设计作品
DADA 2013
STUDENTS

徐卫国 / 尼尔·林奇（英）编
Xu Weiguo / Neil Leach [eds.]

中国建筑工业出版社
CHINA ARCHITECTURE & BUILDING PRESS

目录 / CONTENTS

China

前言 / PREFACE

本作品集收录了“设计智能：高级计算性建筑生形研究”学生作品展中的作品。该作品展作为 DADA 2013 系列展的一部分，将会在 751 D-Park 时尚设计广场举办。自从 2004 年以来，徐卫国与 Neil Leach 为北京建筑双年展合作策展了一系列的展览。第一次是 2004 年在 UHN 国际村举办的“快进 >>”展，之后三次分别是 2006 年在世纪坛举办的“涌现”展，2008 年在 798 时态空间举办的“数字建构”展，以及 2010 年同样在 798 时态空间举办的“数字现实”展。

本次主题“设计智能：高级计算性建筑生形研究”展示了在建筑教育领域中的一系列新进展。第一，“智能”指的是在设计加工领域中越来越普及的智能系统以及新型材料的使用；第二，“高级”一词指在例如 Grasshopper 等的算法软件以及三维打印这样的加工技术已经普及的时代中，逐渐成熟的计算机技术和技能的应用；第三，“研究”指的是在许多研究生课程近年来明显的转变：即由主要基于设计的硕士研究生课程向基于研究的课程设计转变。这些课程往往为之后的博士研究铺设道路。

This is a catalogue of the works on display in the 'Design Intelligence: Advanced Computational Research' exhibition of students' work at 751 D-Park, Beijing, as part of the DADA 2013 series of exhibitions. Since 2004 Xu Weiguo and Neil Leach have been collaborating on a series of exhibitions for the Architecture Biennial Beijing. The first exhibition, 'Fast Forward >>', took place in UHN in 2004. This was followed by three further exhibitions, 'Emerging Talents, Emerging Technologies' in the Millennium Museum in 2006, 'Immaterial Processes: New Digital Techniques for Architecture' in 798 Space in 2008, and 'Machinic Processes' in 798 Space in 2010. This exhibition is a continuation of that collaboration.

The title, 'Design Intelligence: Advanced Computational Research', signals a series of new developments within architectural education. Firstly, the term 'Intelligence' refers to the use of increasing use of intelligent systems and smart materials in the design and construction of buildings. Secondly, the term 'Advanced' refers to the increasingly sophisticated use of computational techniques and technologies, in an age when the use of algorithmic software programs such as Grasshopper, and fabrication processes such as 3-D printing have become commonplace. And, thirdly, the term 'Research' refers to a shift evident in many postgraduate educational programs away from the once dominant culture of design master classes towards a research based agenda that often lays the foundations for PhD research.

本作品集为世界上顶尖的建筑院校的先锋数字设计作品提供了一个展示的平台。包括英国建筑联盟建筑学院，美国南加州大学，哈佛大学，美国南加州建筑学院，同济大学和清华大学等。本作品集收录的是将在 751 D-Park 举办的系列展览的一部分。这些展览包括由宋刚和 Philip Yuan 所策展的装置展，高岩、王鹿鸣策展的中国建筑师作品展，以及由 Patrik Schumacher 所策展的大师展，其参与者包括扎哈·哈迪德事务所，Reiser& Umemoto，让·努维尔，蓝天组，Morphosis，Gehry and Partners。

展览的举办离不开许多个人和组织的帮助。特别感谢 751 D-Park 时尚设计广场为展览提供的场地支持。

我们十分感谢为作品集编排和设计付出努力的所有人。特别感谢：金漪、李晓岸、刘傲、刘春、吕帅、娄晓一、林秋达、沈源、王捷、翟炳博、罗丹……为作品集的编排、设计和翻译所做的贡献，以及为作品集贡献材料的各个学校。

徐卫国

尼尔·林奇

The catalogue offers a showcase of the most advanced computational design work from some of the leading schools of architecture in the world. These schools include the Architectural Association, University of Southern California, Harvard GSD, SCI - Arc, Tongji University and Tsinghua University, etc. The work included here is part of a larger exhibition being held in 751 D - Park, Beijing, that includes a series of installations curated by Philip Yuan, Gang Song, and an exhibition of some of the most famous architects in the world - including Zaha Hadid Architects, Reiser & Umemoto, Jean Nouvel, Co (op) Himmelb (l) au, Morphosis, Gehry and Partners - and curated by Patrik Schumacher.

The organizers are grateful to all those who have contributed to the preparation of this catalogue. In particular, they would like to thank Yi Jin, Xiaoan Li, Ao Liu, Chun Liu, Shuai Lv, Xiaoyi Lou, Qiuda Lin, Yuan Shen, Jie Wang, Bingbo Zhai, Dan Luo for their invaluable contribution in helping to design, translate and compile material for this catalogue, and to the schools themselves for submitting that material.

Xu Weiguo

Neil Leach

对话 / DIALOGUE

徐卫国–尼尔·林奇
XWG-Neil Leach

徐卫国（以下简称徐）：你对全球蓬勃发展的数字建筑教育怎么看？

Neil Leach（以下简称 NL）：我认为，我们处于一个对于建筑院校的数字化教育而言非常重要的时间点。对于中国尤为如此。在我 2004 年第一次来到北京的时候，没有人知道什么是编程。而现在种种数字化研习班层出不穷，包括现在清华大学正在举办的这个。在为这种变革欣喜之余，我们也需要面对一些新的问题。由于当下，数字化已经逐渐普及，所有人或者说绝大多数人会使用 Grasshopper。那么在当今环境下，数字化到底扮演着什么角色？数字化展览的定位究竟是什么？因此我认为，这就是为什么"高级数字化"一词必须囊括在本次展览的标题之中。因为如果我们想要有重大的突破和发现，我们必须找到一些新的革命性的数字技术应用方式。

徐：你提到了"高级的"一词，你认为在数字建筑技术领域，"高级的"数字建筑技术主要是指哪些方面？

NL：在我的概念中，这包含两个方面。其一是在设计过程中高级软件技术的应用，而高级数字化技术的另一方面是在制造中的应用。对于软件技术而言，我们可以区分诸如 Rhino 和 Maya 这样随处可见的特定软件，越发流行的算法工具例如 Grasshopper 和 Processing，和像 Catia 和 Digital Project 这样的参数化工具。持续处于高级领域的是代码的使用，即编程或者算法。换句话说，即生成的算法手段等等。当然，所有这些每一天都在进步着。还有许多 Grasshopper 的插件也变得越来越专业化。

另外一个方面是数字化生产。如同我们可以在许多学校（例如 AA, Bartlett，清华，同济，IaaC，SCI-Arc，ETH，南加州大学，密歇根大学）里面看到的那样，他们现在都在投资机器人。事实上，许多学校都在尝试利用机器人的生产模式。目前，三维打印并不意味着什么，因为所有人都在使用它。这并不是高级的。但是机器人是，以及交互工具的使用也变得越发重要起来。在中国，这方面越来越流行。因为 Arduino 模块以及 Servo 在这里均非常便宜。我注意到在例如 SCUT 和清华这类学校，许多工作精力都投入交互设计的领域。在 USC，DIA 以及 Bartlett 中，我们也在使用 Kinet，记忆纤维等等。许多不同的技术在被使用。世界自 2004 年我们举办第一个展览以

XWG: The title of the catalog is Design Intelligence: Advanced Computational Research. Where do you think digital education is leading?

NL: Well, I think that we are at an interesting moment in terms of the use of computation in schools of architecture, especially in China. When I first came to Beijing in 2004, nobody knew what scripting was. Now there has been this revolution. There are many computation workshops including the one here in Tsinghua University. But now we have a different issue, because now computation is everywhere. Everybody - almost everybody-can use Grasshopper. So, what is the role of computation today? What is the role of an exhibition that is looking at computation? So I think that is why the term 'advanced computation' has to be in the title of this exhibition, because if we want to find something significant, we have to find some new revolutionary, progressive use of computation.

XWG: When you use the word 'advanced', what do you mean in relation to digital technology?

NL: To my mind, there are two aspects. One is the focus on the use of advanced software techniques in the design process, and the other side is advanced computational technologies in fabrication. In terms of software techniques, we can distinguish between the use of explicit modeling tools that Rhino and Maya that are used everywhere, algorithmic tools such as Grasshopper and Processing that are becoming extremely prevalent, and also parametric tools like Catia and Digital Project. What is advanced and remains advanced is the use of code - the use of scripting or algorithms, in other words-and especially the use of techniques such as genetic algorithms and so on. And of course these are getting more advanced every day. There are also a lot of plug-ins for Grasshopper that are getting more and more specialized.

The other aspect is digital fabrication. And we can see that many schools-such as the AA, Bartlett, Tsinghao, Tongji, IaaC, SCI-Arc, ETH Zurich, USC, Michigan and so on-have now invested in robots. In fact many schools are looking at robotic fabrication. And so simple 3D printing means very little these days, because everyone is using it. It's not advanced. But using a robot is. And the use of interactive tools has become increasingly significant also. In China these have become especially popular, because Arduino boards and Servos are extremely cheap. I noticed example, especially in schools such as SCUT and Tsinghua, a lot of work is going on with interactive technologies. And in schools like USC, DIA and the Bartlett we are also using Kinect and Shape Memory Alloys; so many new technologies are being used. The world has progressed a lot since 2004, when we held our first exhibition.

来有了大幅度的进步。

徐：那你是不是认为我们的数字建筑教育上了一个台阶？

NL: 绝对的。很明显自从 2004 年以来有很大的变化。事实上，我的预测是在这个十年，即 2020 年结束的时候，我们甚至不会再提“计算机辅助”这个单词。因为它将涵盖所有方面，以至于没有哪里是特殊需要专门被提出的。因此再讨论这个是没有意义的。就像在施工现场，我们不再讨论电动设备因为所有的设备都是电动的。

徐：这种在建筑教育上的变化对将来的建筑实践意味着什么？

NL: 事实上，建筑实践与建筑教育之间的关系从来都是一个有趣的课题。在许多方面，我认为建筑实践已经走到了建筑教育的前面。在中国的某些学校，我听说当学生使用计算机工具的时候他们的教授说：“不要使用这些工具，我们应当使用包豪斯的方式。”事实上，我正在包豪斯任教。那里我们对于计算机技术的使用已经达到了一个非常高的程度。许多学校的老师和教授依然对于新的计算机工具的使用持保守态度。但是，让我们看看最好的那些建筑公司，例如盖里、福斯特、扎哈·哈迪德的事务所，在其中我们会发现它们都有研究小组。他们正在研发计算机工具以帮助今天的复杂建筑得以在预算以及限定时间内实现。在盖里的事务所中有盖里技术公司；在诺曼·福斯特事务所中有专业建模小组；在扎哈·哈迪德事务所中有 CODE 等等。当你咨询商业公司时，他们也会告诉你他们需要会使用 Grasshopper 的学生。但是还是有很多学校连 Grasshopper 都不教授。

徐：我同意你的观点，现代的建筑实践在某些方面是领先于建筑教育的，但是实际上，在建筑学院里面的教学和研究，比如在软件程序方面和构件加工方面的研究，是实践单位及实践事务所所无法做到的。比如说，在清华，我们多年以来从事的对于建筑形态的研究，是通过实验来找到平常不存在的动态的形象，这样做拓展了建筑的形态领域。因为从传统的建筑设计的角度来看，那些基于几何的形态数量是有限的；那么如何发现一些新的动态的形象，就需要通过研究来发现，如果没有这些研究，那些几何的形态会被用得越来越少。通过学校的 studio 对形态的发现会给建筑带来更广阔的景象，

XWG: Do you mean that digital education has progressed to a new level?

NL: Absolutely. Obviously things have changed since 2004. Actually, my prediction is that by the end of this decade - by 2020 - we won't even use the word 'computation', because it will be everywhere - everywhere and nowhere. So it would be pointless using the term. Just as on the construction site, we don't talk about electrical tools because everything is electrical.

XWG: How will this shift in architectural education impact on architectural practice in the future?

NL: Well, in fact, the relationship between architecture practice and education is an interesting one, because in many ways I believe that architectural practice is often more advanced than education. In some schools in China, for example, I hear that when students use computational tools, their professors tell them: 'No, no. Don't use these tools. You should use the techniques that they use in the Bauhaus.' Actually I teach in the Bauhaus, and there we use these computational techniques at an often very advanced level. So we often find in schools of architecture that the professors are reluctant to open up to the use of new computational tools. Meanwhile, when we look at the leading architecture practices, such as the offices of architects like Frank Gehry, Foster, Zaha Hadid, we will find research groups in each of these offices, where they are developing computational tools that are necessary in order to get the complex buildings of today built on time and on budget. So in Gehry's office you find Gehry Technologies; within Norman Foster's office is the Specialist Modeling Group; within Zaha's office CODE, and so on. When you also talk to commercial firms, they will tell you that they need students that can all use Grasshopper. But there are still schools in the world that don't even teach Grasshopper.

XWG: I agree with your comments that the current architectural practice is more advanced than architectural education. But, in fact, within the curriculum of the department of architecture, research on software and programming and fabrication technology is beyond the capability of practicing companies. In Tsinghua for example, within the past several years, we have done a number of experiments based on formal studies. Through experiments, we try to capture the dynamic forms that do not exist in normal practice. Thus in this way the realm of architectural geometry is expanded. In traditional architectural design, there are only limited forms based on rationalized geometry. It is worth researching how to discover new dynamic forms. Without this type of research, we would exhaust the current library of geometrical forms. Through the studio experiment within the school, we may enlarge the perspective of architectural design. This is one aspect that practicing architects are not capable of.

这些是那些实践单位没有办法实现的。另外一个方面，就是你刚才所谈到的关于高级技术用于加工领域的问题。其实，现在各建筑学院，包括西方的建筑学院，例如：ETH、哈佛 GSD、AA School 还有中国的华南理工大学、清华大学等做了很多关于互动建筑的研究。这些研究完全是探索性的，使用了高级技术来操控建筑的构件从而形成互动或环境响应装置。这种装置它不具有实际建筑的功能，但是它是一种探索，为将来建筑有可能变成互动建筑提供了基础。同样，在这点上，实践单位及建筑事务所也是做不到的。

NL：我完全同意你的评论。我认为在计算机革命之后的二次革命将会在教育领域发生。这并非一个全球化的过程，但是已经露出了苗头。我将设计课视为一种实验室，一个探索性的实验室。这个概念起源于屈米于 1990 年代在哥伦比亚大学创立的模式，他们往往将设计课称为实验室。但是我认为这个概念被例如 AA DRL 这样的项目带到了一个高度。在那之后这种模式被 Achim Menges 及他所在的斯图加特的电脑设计学院（ICD）带上了另外一个层次。在那里，他非常严谨地面对研究问题。因此，在我的设想中，这次展览的参展学校将从多方面反映此变化。

这是一种由传统老师—学生式的设计课程向新的扎根于研究的工作模式的转变，新的模式往往关注于材料以及计算机的研究。我认为这是一个重要的转变。在我的设想中，我们应当重新审视我们的硕士教育：它应该逐渐由一个单纯的本科教育的延伸，向为博士做准备的研究导向性的项目转变（同时我认识到博士项目也成为了另外一个严峻的问题，因为世界上许多地区仅仅接受具有博士文凭的教授）。一个非常重要的进展是我们在建筑院校中试图建立一种博士文化，换句话而言，硕士项目对于变革我们对研究生工作的认识非常重要，它定位于将设计课转变为研究型实验室，并为博士课程做好准备，从而提供有博士学位的专业学术人才。

徐：是这样的，学校里的设计课程确实是作为一个实验室在创造着新的知识。那么，你认为学校课程所创造出来的新知识将怎样有益于建筑实践及影响未来建筑设计？

NL：我们可以观察到在实践事务所中有一部分研究实验室在进行软

The second aspect is, as you've just mentioned, the application of high-tech architectural fabrication. In fact, within many architecture schools especially in the West such as ETH, Harvard GSD, AA, or in some schools in China, such as SCUT and Tsinghua, there is a lot of research being done towards interactive architecture. This purely exploratory research uses high - tech driven architectural components to form interactive installations. These installations don't have the function of real architecture. But they are the foundation of a potential interactive architecture of the future. Similarly, this is also another undertaking that cannot be done in the architecture practice or firms.

NL: I agree with your comments. I think that the second revolution happening beyond the computational revolution is the revolution in education that is beginning to appear. This is by no means universal, but it is beginning to appear. And I see the design studio as a kind of laboratory, an experimental laboratory. That started maybe with people like Bernard Tschumi in Columbia in the 1990's. He always referred to the studio as a laboratory. But I think this has been taken to another level by programs such as the AA Design Research Laboratory(DRL). And since then this in turn has been taken to another level by people like Achim Menges in his Institute of Computational Design (ICD) in Stuttgart where he is very rigorous in the way that he approaches the question of research. So, to my mind, this is reflected in many ways by some of the schools in this exhibition.

There is a shift from the old fashion design studio, which is only about design and the traditional master-student relationship, towards a new kind of project which is fundamentally grounded in research, often in terms of both material fabrication and computational research. That I think in an important shift. In my mind, it's a shift in re-envisioning the master's program especially from simply being a continuation of the undergraduate education towards one with a greater research orientation that lays the foundations for PhD research. And I think that PhD research is another issue that is becoming really important, because many countries in the world will now only accept professors who have PhDs. One of the important developments has therefore been trying to build up a PhD culture in schools of architecture. So, in other words, it seems to me the master's programs are really important in revolutionizing our understanding of what postgraduate work should be - in changing the design studio into a research laboratory and then feeding into a PhD program in order to produce professional academics with PhD degrees.

XWG: The design studio acting as a laboratory is generating new knowledge. What is the benefit of this knowledge generated in the school to professional practice, and how might it even further influence architectural design in the future?

件研发，甚至研究材料特性，同时在建筑学校中也有。这两者之间具有沟通交流的桥梁。在许多案例中，曾经就读于例如 AA EmTech 硕士项目的学生很大比例进入了实践事务所的研究室里。由此，两者之间存在一定沟通和联系。

同样的，我们也可以看到那些与学校教育有直接联系的实践事务所，最好的一个例子就是 AA DRL 项目和扎哈·哈迪德事务所。许多学生在 AA DRL 学到必须的技能后去扎哈·哈迪德这一类的事务所工作。这里有一个直接向特定事务所提供具有技能的毕业生的过程。因此，我认为我之前观察到的，建筑实践领先于建筑教育的趋势有一个潜在的转变过程，逐渐变为它原本就应当成为的模式。我们在过去曾见过这个模式，如 BIM 最初起源于 MIT 这一类的教育机构，但它在实践中发展。因此，我们看到了一个更为自然的模式。即理念最初在学校中探讨，之后通过建筑实践发展成熟。

徐：比如用互动建筑作为例子。实际上，让建筑能动，让建筑能够适应于环境，或者说建筑能够随着人的活动自调节自适应，可以说这是人类自古以来的理想。谈到建筑能够适应人类、适应自然的要求，其实在建筑的历史上有很多的探索。例如赖特的有机建筑，还有布鲁斯·高夫的连续的建筑，再比如巴特．普林斯直接模仿动物形态的建筑，也有以生物习性为基础来设计建筑的例子，这些努力都试图让建筑融入环境，其目的都是希望建筑与环境互动，创造更宜居的居住环境。可是在过去，受限于技术，他们只能做到一定的程度，设计受到了局限。但是现在，比如说建筑学院对于互动式建筑的研究，这些技术如果成熟的话，它将真正意义上实现建筑与环境之间的互动。从而实现历史上建筑师乃至人类对于动态建筑的崇高理想。从这个角度来说，建筑学院里的研究可能会导致将来建筑变成环境中真正的一个生物，就像人一样，可以活动。

NL：是的，我同意。我认为问题是计算机如何能帮助我们实现那个梦想。此时很重要的是我们需要回顾一下计算机发展的历史阶段。我认为在 1990 年，那时我们几乎是对于虚拟现实的一种类似于科幻小说一样的认识。人们想象着未来的可能性却又不了解它们到底是如何实现的。之后，进入了另外一个我称作“数字建构”的时代，

NL: We can observe that there are research laboratories both inside architectural practices - developing software, and in some cases even working on materials, material behaviors - and also in architectural schools. There is dialog emerging between the two. In many cases, students who have been working in the master's programs such as the EmTech program in AA then go to work in research laboratories in architectural practices. So there is obviously some kind of dialog going on. And we can probably distinguish between practices that are connected to education, and other practices that are not. The best example of the former might be the relationship between the AA DRL and Zaha Hadid Architects (ZHA). Many of the students develop techniques and skills while studying on the AA DRL program and then go on to work in offices such as ZHA, so that there is a continuation and a direct way of feeding certain architectural offices with skilled graduates. So I think there is a potential of reversing what I have seen in the trend recently, where practice has been ahead of education, and getting back to what it should be. I think this has been seen in the past, where, for example, the development of Building Information Modeling came out of work at academic institutions such as MIT, so that we are getting back to a relationship that's much more natural, where the idea is first explored in schools, and then becomes matured in its application within professional practice.

XWG: Let's take interactive architectural research, for example. My own opinion about interactive architecture is that it will have a big influence on future architecture. Fundamentally speaking, the goal of interactive architecture is to create architecture that could adapt to the environment, or, in another sense, architecture that could self organize according to human activities. This has been the fantasy of human since ancient times. Speaking of how architecture would accommodate human needs, there have been many explorations in architectural history such as Frank Lloyd Wright, Bruce Goff, and other designers producing Organic Architecture. Some of the designs are the direct mimicking of organic forms from nature, while some are simulating the habit of living creatures as the foundation of architectural design and so on. These are all fantasies aimed at creating interaction between architecture, environment and people, generating better living conditions. However, there were only certain possibilities in the past as they were limited by technologies. If the technology that they are being developed currently within interactive research teams in design institutes were to mature in the near future, it would achieve the fundamental interaction between architecture and human beings. It would realize the dream of a dynamic architecture that we have dreamt of since ancient times.

NL: Yes, I agree. I think that the question is really how computation helps that dream. I think

我们有史以来第一次开始利用计算机来了解结构和材质特性。现在，我们进入了第三个阶段，此时计算机的使用成为必须的。它不仅仅可以帮助我们了解材质特性，更有助于我们控制建造的过程。如此，我认为这里存在三个明显的阶段，同时在电脑控制的交互设计中我们也可以看到这三个明显的阶段。

在我的概念中，许多 1990 年代的交互式设计工作仍然局限于科幻世界中。人们有一个建筑可以适应并且改变行为和外形的意识，但是没有人知道它应当如何实现。最成功的早期试验的范例有例如 Mark Goulthorpe 的 Aegis Wall 项目。在这个项目中他与 RMIT 建筑空间信息实验室（SIAL）的 Mark Burry 合作开发了一个可以移动的表面。它在很多方面是个非常原始的原型，但却是一个重要的开端。现在突然之间，我们拥有许多现成的廉价技术，至少在中国它们很廉价，就如同 Arduino，Servos 等等，使得这一切价格合理、可行。我自己的看法是最终这一切将从目前的几乎是对技术的迷恋开始进化，即目前人们仅仅将特殊的技术制作成艺术装置来展现我们到底能做什么，就像当我们最初开始使用机器人以及数字加工技术时也有对于技术本身的迷恋。但是在我们自己的展览目录中，David Ruy，2010 年东海岸区域的策展人，有一句评论："我们现在进入到了一个并不仅仅是对于技术本身而盲目迷恋的阶段，而是我们应当如何在日常的设计中创造一种新型的敏感又精致的设计方式。这就几乎已经超越了单纯对于技术的迷恋，而致力于产出更加精妙的设计，这些技术将像假肢一样帮助人们行动和生活。"

对于交互技术而言，同样的事情也会发生。就目前而言我们大量的精力投入于开发我们到底能做什么，以及这个技术本身看上去非常特别。但是我认为技术逐渐会变成日常生活的一部分。交互设备将以非常实际的方式被使用。我们不会将它们变成艺术装置，它们会变成建筑语汇的纯粹的一部分，就像计算机已经成为我们工业设计的一部分一样。Perry Hoberman，我在南加州大学的交互媒体的同事，曾经说过：在未来，交互技术将变得如此常见以至于我们不会注意到他们，例如有时当你去洗手间时，灯会自动亮起，就是这么简单。我认为我们将会有对于环境反馈的智能建筑，甚至会出现对我们的

it's important maybe to look into the history of computation in terms of different phases. I like to think of what was happening in 1990's was when there was an almost science fiction like understanding of what they called 'virtual reality'. People were dreaming about a possible world without really understanding it. And then, the next phase is what I call 'digital tectonics' - when for the first time we started using computers to understand structural and material behavior. And now we have entered a third phase where it's absolutely necessary to use computation not only to understand material behaviors but also to control the whole building process. So I think there are three distinct phases. I think you can also begin to see computational driven interactive architecture in terms of these three different phases.

In my view much of the work done in the 1990s about the possibilities of an interactive architecture was still locked in to this world of science fiction. People had this idea that the architecture could adapt and could change its behavior and its form, but nobody knew how to materialize it. The most successful of the early experiments were examples such as Mark Goulthorpe's Aegis Wall project where he collaborated with Mark Burry of the Spatial Information Architecture Laboratory (SIAL) at RMIT developing a surface that would move. That was primitive in many ways, but it was an important start. Now all of a sudden we have readily available technologies that are extremely cheap - cheap in China anyway - such as Arduino, servos, and so on, making all of this possible and economically viable. My own view is that eventually this is going to evolve from its present state which is really almost like a fetishization of these technologies - with people making art installations showing what you can do, simply because it seems so special. In a way in the beginning when they first used robots and digital fabrication technologies there was a kind of fetishization of the technologies themselves. But in our own exhibition catalogue there was a comment made by David Ruy, the curator of the East Coast section in 2010, that we have now entered a new stage where it's no longer a question of simply fetishizing these technologies, but about how we might use them as part of everyday design, and create new sensitive, exquisite forms of design. So it is almost as if you transcend that sort of obsession with the technology and produce something exquisite as though these technologies simply become a prosthesis to human operations.

I think that the same is going to happen with interactive technologies. At the moment we are heavily involved in what we can do and it seems very special. But I think eventually it will become part of everyday life. Interactive devices will be used in very pragmatic ways, and

思维有所反馈的智能建筑。

目前，我们已经有了一部分和神经科学家的合作研究。此类研究关注于我们如何使得建筑与我们的思维交互，所有的东西都将是自动化的。因此，我认为这是真正的交互智能环境的未来。“智能”这个单词非常重要。因为我们有智能材质，智能建筑，在某方面智能建筑将向一个几乎不会被主观意识到的方向前进，这些系统将融入在标准的建筑设计和施工中，我们甚至不会注意到他们的存在。

徐：让我们来讨论一下互动式建筑的探索都带来了什么变化？首先，智能材料能够随着周围变化的环境而发生相应的变化；其次，建筑节点也可以改变其形状来为建筑内部及外部的人们提供更好的条件。还有其他的方面吗？

NL：我认为这就是最重要的两点。

徐：当然，建筑师王弄极所设计的长沙中联重科展示中心就是采用了互动式建筑设计手法，建筑能够智能地打开建筑的窗户甚至屋顶，使整个建筑的形体产生变化。同时，对于室内的空间，有一些构件、墙、顶或地板可以动，动完之后可以重新划分出新的空间。或许这是互动建筑带来的第三个方面的变化，即可以重组空间。

NL: 许多情况下这最终归结为一个尺度的问题。例如，当空调最早被发明时，它非常的昂贵，现在当我们有许多人使用空调时它变得廉价。对于手机也是如此，当使用者越来越多时，它的价格将下降。就我的认识而言，目前 Arduino 非常廉价，它也是开源的，这样人们可以在所有的地方小尺度地使用它，因为我们目前还没有相应的技术大尺度地使用它，但是很明显，这终将发生。我完全同意你的观点，即两个交互方向是对于环境的反馈以及对于使用者的反馈。我们就如何达到此目的的探索仍然在初级阶段，但是它一定会发展，因为它将是建筑的一个重要的待发展领域。

徐：这个领域的发展对建筑师来说意味着什么呢？对于建筑师的设计来说，是不是建筑师的设计内容会发生根本性的变化？

NL: 我认为这已经为建筑设计领域带来了根本性的变化，我们可以从建筑设计的过程中看到这个转变。过去人们认为建筑师是一个由

we won't make art installations about them. They will simply become part of language of architecture in a way that computation has become part of how we produce designs today. Perry Hoberman, one of my colleagues in USC who works in interactive media, once said that in the future interactive technology will be so commonplace that we won't even notice it. Sometimes when you go into a toilet, the light comes on automatically when you close the door. It's almost as simple as that. I think we will have intelligent buildings that will respond environmentally. And you'll have intelligent buildings that respond to the way we are thinking. There is already research going on with neuroscientists looking at how we can make buildings interactive and connected to the way we think. Everything will be automatic. So I think that's the real future of responsive and intelligent environments. The word 'intelligent' is really important. Because just as we have smart materials, we also have smart or intelligent buildings. And we have aspects of intelligent buildings that are developing in such a way that they will soon become almost invisible. These systems will be embedded in standard building design and construction so that we will not even really notice it.

XWG: Let's look at what has happened in interactive design research. Firstly, with regard to the materials, intelligent materials can adapt to the nature or environment. Secondly, the compartment of the building could change its form to provide better conditions for the people both inside and outside. Is there any other type?

NL: I think these are the two important ones.

XWG: Of course, I think that one example is the building designed by Wang Nongji in Changsha, which has a mechanical system that can open the wall and roof. Maybe that could be the third aspect of the interactive design, where a building could move its components at a large scale. Or with interactive interiors, elements such as doors and walls could move creating new programmatic space.

NL: I think that often it comes down to a question of scale. For example, when air conditioning was first developed, it was very expensive. Now you have many people using air conditioning, so that it has become cheaper. The same is true for cell phones. Once you have many people using them, they become cheaper. As far as I'm aware, at the moment - although Arduino is really cheap, as it is open source, so that people have been using it everywhere at a small scale - we still don't have this kind of technology readily available at a building scale. But obviously, that is going to happen eventually. I think you are obviously right in saying that the

上而下的设计者，控制着所有的东西，但是现在建筑师逐渐转变为一个由下而上的设计者。例如，当我们写一个能生成设计的算法时，这与传统的通过草图来设计的方式有着极大的不同。所谓的交互建筑也有类似的元素，除了设计形体，换句话说，除了通过形体逻辑或者风格的分析来设计，我们现在在设计行为、过程，或者措施，最终导致形体的生成。由此建筑本质将发生改变，就像设计过程的本质也在改变一样，它将从根本上为我们如何设计带来变革。

徐：这个是不是说明建筑师设计的内容变化了，同时，未来设计的方法也将发生彻底的变化？从今天的角度看，是不是我们的建筑教育也要发生根本的变化？

NL: 我认为最根本的改变从后现代主义的关注点上转移开，从对于风格和表现的问题上转移开来，而转向对于过程和材质的讨论。这个改变并不仅仅是在设计工作室和实验室中发生，它还影响到整个的理论框架。建筑设计的基础理论已经发生了变革，我们可以看到 Manuel De Landad 的工作所发生的改变，他是一个关注于材料特性以及表现的哲学家，而非传统的专注于研究象征手法以及语义。我们通过一些滤镜，例如哲学家 Jacques Derrida 可以看到我们对于建筑学认知的转变，他关注于通过全新的视角即我们如何通过过程和材料特性诠释世界。在某些方面，这让我想起马克思的一句评语：“目前为止，哲学家仅仅诠释了世界，而真正的挑战是改变世界”。因此，在教育事业中，我认为在设计课和历史讲座中应当改变讨论重点至“应当是什么样”，以及“我们如何看待建筑的材质性”。由此，我们可以抛弃后现代主义对于建筑风格的关注等。我们需要为建筑学院中的理念带来变革。

徐：传统的建筑教育主要训练学生们对于形态、空间、环境、材料、结构等这些方面的设计能力，而新的建筑教育可能会更多地关注行为、智能化等这种新的知识。你认为还有什么需要发生变化呢？例如是否会有新的设计哲学、设计理论等？

NL: 我认为我们思考建筑的方式在设计哲学和建筑理论方面已经完全改变了。在我的脑海中，当讲到设计理论时，我们必须问自己设计理论究竟是什么。在某些方面，Achim Menges 所做的利用科学与

two areas are environmental responsiveness and responsiveness to the user. We are still at an early stage exploring how we can do that. But for sure it will develop, because that seems to be one of the important areas where architecture needs to develop.

XWG: What do you think this could possibly mean for architects and their designs? Will there be a fundamental change in the content of architectural design?

NL: I think that it has already caused a fundamental change in architectural design. We can see this from the way in which the process of designing buildings has changed. The old fashion idea that the architect is a top down designer, controlling everything, has given way to the notion of the architect as the designer of bottom - up processes. So for example, when you write an algorithm that generates a design, it's a very different way of designing a building to the old fashioned way of sketching something out. So what this means for interactive architecture is something very similar. Instead of designing forms - or in other words, instead of designing something according to the logic of form, or the logic of style - we are now designing behaviors, or processes, or operations that will lead to the generation of forms. So the very nature of architecture is going to change, just at the very nature of the process of design is also changing. It's going to fundamentally revolutionize what and how we design.

XWG: So this proves that the content and the method of architectural design would fundamentally change. But let's go back to the topic of architectural education. In the current situation, how should we adjust architectural education to take account of this change?

NL: I think that at a fundamental level the shift is away from the concerns of Postmodernism - concerns about questions of style and representation - towards an interest in processes and material behaviors. And this shift operates not only in the laboratory of the design studio; it also affects the theoretical framework. We have to revolutionize the kind of thinking that underpins architectural design. I can see that shift already in the work of Manuel De Landa, who is a philosopher interested in material behaviors and expressions instead of the old fashioned obsession with symbolic meaning and interpretation. I can see a shift away from the way we think about architecture - through the lens, say, of philosophers such as Jacques Derrida, who are interested in interpreting the world - to a very different idea of thinking how we can change the world through processes and material behavior. So in some ways it reminds me of the comment Karl Marx once made: "Up until now, philosophers have merely interpreted the world. The point is to change it." And so in education I think there is a fundamental shift both in the design studio and in the theory seminars to what it should be, about

材料特性的工作已经是理论的一种了。我们注意到了我们可以开始讲科学和技术理论化。由此，在未来我们的操作手法也会改变。对于我自己的学生而言，这个变化已经开始发生了，不像过去写一些关于理论的论文，他们已经开始就“材料哲学”方面做了一些研究，他们的研究工作是通过材料试验，写软件等来思考建筑。因此，设计和理论的学术本质将会发生改变，在某些程度上，它们已经发生了改变。

徐：我同意你的观点。或许教育的方式应该发生变化，不应该教学生应该怎么做设计，而是应该更多地关注研究，应该教学生获取新知识的方法，将设计课程从一个教学的课程变成一个研究的课程。

NL：我认为建筑教育最重要的变化是我们取得知识的方式。越来越多的，我们开始利用网络来寻找信息。事实上，让我们想一想大学的图书馆。现在很少人真正地使用图书馆，除了把它当做一个安静的自习室。因为我们可以轻易地在网上买到更便宜的书而且不必担心为逾期的图书缴纳高昂的罚款等等。最重要的是，人们寻找信息的习惯也已改变。当我们想要寻找资讯的时候，至少就我个人而言，当我想写点东西时，第一件事是上谷歌或者其他的搜索引擎以寻找网上的信息。图书出版业面临着一个危机，因为人们不再买那么多的书而转向其他获得信息的方式。就技术而言，也面对着类似的问题，学生逐渐不再必须向老师寻求知识的传授，而转向在线培训视频。例如 Josa Sanchez，他马上要开始在 USC 任教，但是全球已经有至少 50 万学生在使用他的在线 Processing 教程。在这个意义上，我们涉及的是全球化的课堂，所有人可以看到其他大学的网络课程只要他们可以进入 Vimeo，Youtube，或者它的中文版。所有人可以从网上看到其他大学的参考书目以及课件，这对于建筑教育的结构而言是一个挑战，或许在未来，教授的角色不再是教导学生而是简单地教学生如何寻找信息。因此，未来的教育结构也会发生改变。

徐：你是否认为图书馆会消失呢？

NL: 图书馆会变成咖啡馆或者酒吧这样的地方。人们可以去那里喝个咖啡之类。

徐：这是个有趣的观点。如今发展非常迅速，未来将会有许多新功

how we approach the actual materiality of architecture. Therefore, we should leave behind the concerns of Postmodernism about 'styles' of architecture, and so on. We need to revolutionize the whole way of thinking in schools of architecture.

XWG: Traditional education trains students in areas such as form, space, environment, materials, structure, lines, and so on. But maybe architectural education should now extend to include behaviors, intelligent technologies etc. What new areas do you think that it is necessary to include?

NL: I think that the way we think about buildings has already completely changed in terms of design philosophy and theory of architecture. To my mind, when talking about design theory, we have to ask what design theory is. In some ways, what Achim Menges does in terms of working with science and materials is actually a form of theory. What we notice is that we can now begin to theorize the world of science and technology. So the way we operate in the future might well change. With my own students what is already happening is that, instead of writing essays about theory (in the old fashioned sense), they are already doing research about 'material philosophy'. Their research is about thinking about architecture through material experiments, writing software and so on. So the whole nature of how academics approach design and theory will change and - to some extent - has already changed.

XWG: I agree, so maybe the educational method has to change. We should recommend a kind of research based teaching method. Now we are only teaching students what they should do, but what we should really teach is how to obtain new knowledge. So we should change the studio teaching into research based teaching, trying to discover new things for the future.

NL: I think that the other important way education is changing is in how we access knowledge. Because increasingly we are using the internet to find information. In fact, when you think of a university library, very few people use the library these days except as a quiet place to work because they can get much cheaper book online where they don't have to pay fines for the overdue books and so on. And, moreover, they access information in a different way. The first thing they do - and indeed the first thing I do when I try to write something - is to access Google or other search engines to find out information online. There is obvious a crisis in the production of books because people are no longer buying so many books, and are finding other sources of information. The same goes for the technical side of thing. Increasingly students are finding out information not necessarily from their professor, but through some online tutorial. For example, Jose Sanchez, who is about to start teaching in USC, already has half a million students around

能的建筑出现，也很有可能这些已经存在的建筑会变成其他的功能。前段时间，清华设计课程有一个很有意思的设计，要求学生们调查北京的新生活方式，他们发现，如今北京的购物模式发生了变化，越来越多的人都会选择网上订购商品，因为网上的商品会相对便宜，但是人们有时候发现收到的商品与自己所预期的相差较大，于是，他们希望能够见到商品的实体甚至可以对商品进行试穿或试用。针对这种情况，许多人选择去商品实体店先看一下商品本身，然后再到网上进行购买。于是，这些商店变成了仅仅是展示商品的用途。这是一种很有趣的新生活方式，针对这种方式，学生们进行了商店的设计，白天的时候，商店作为商品展示和服装试穿的功能存在；到夜晚的时候，商店空间用作旅馆，商店的中心变成了旅馆的门厅，一个个试衣间变成了旅馆的客房，并且每个客房针对不同的需求采用了不同的空间形状。这是一种新型的建筑，功能和形态都可以随着要求进行变化。

NL：其实交互设计还有另外一个方面是形体的朝向与外形会适应内部的功能而产生变化。Greg Lynn 在适应性结构方面已做了一些工作。当然，最有名的是 Rem Koolhaas 所设计的功能可变的 Prada 变形建筑。我认为这方面有许多正在发生的变革，它将影响生活的方方面面。对于购物，库哈斯设计的 Prada 店会存储您的个人信息并且在客人进入商店的时候识别，就和我们网购的时候类似。网站记住我们上次所买的东西并因此给我们推荐一些我们可能喜欢的新商品，例如“买过某本书的人通常也会购买这本书”。由此产生了一种全新的购物结构。事实上曾有这么一部电影在 2002 年上映，即斯皮尔伯格所导演的《少数派报告》，这是一部富含智慧的电影，其中许多并非完全的科学幻想，而是导演试图依据现实信息而预测未来。斯皮尔伯格与 MIT 媒体实验室等专家合作，以预测未来会发生什么，事实上，很多其中的预测被证明是正确的，例如，电影中描述了一种设备会感知手部的运动，这与现在的 Kinect 非常相像。电影预测我们可以根据虹膜识别身份，现在我们在通过移民局的时候也使用相同的技术。同样的，我们也可以进入一个商店时被识别，这些技术为我们生活的各方面带来了变革。

the world following his online Processing tutorials. So in some senses, we are talking about a global classroom. Everyone can see lectures from other universities if they can access Vimeo or YouTube or the Chinese version of that. And everyone can access reading lists and texts and so on from other universities. That is challenging the structure of education. And maybe the role of professors in the future is not to teach people but to show people how to simply access information. So the entire structure of education is changing.

XWG: Do you mean the library will disappear?

NL: Libraries will probably turn into places like coffee bars where people will go and drink coffee.

XWG: Interesting concept. In fact, based on cultural changes, in the future there will be new types of architecture. We should therefore adapt existing buildings to other uses or programs. There are some interesting things happening in the Tsinghua design studios. Some students are now making research into a new type of life in Beijing. For example, the way to shop has changed. Now it's popular to buy things from the internet because it's cheap. But sometimes people get things that don't meet their expectations, so people want to see the items before buying them. Thus there is a new type of shops emerging. These kinds of shops are solely dedicated to showing the product, where people can try out the clothes and then go back to the computer to buy the product. This is a very interesting new lifestyle that influences shop design. This shop is only for showing and experiencing the product in the daytime. And at nighttime, the shop has the potential to be transformed into hotels, especially it has the rings of dressing rooms in the perimeter as trying out is an important program for this shop. In fact this is a brand new type of program. At the same time, all rooms have different spatial qualities to accommodate the need of individual client. Society today is undergoing a rapid transformation, in terms of lifestyle, construction, the way we see buildings and so on.

NL: There's another version of responsive design, where the orientation of the form can change so you can adapt the use. Greg Lynn has some work on this investigating how you might turn or re - orientate a structure in order to change its use. Of course, famously Rem Koolhaas has designed the Prada Transformer building which could also change its function. I think there are many revolutions that are happening. This is affecting everything, every part of life. In terms of shopping, you probably know that Rem Koolhaas's Prada stores recognize who you are when you come into the shop, in the same way when you order online; they'd know what you've ordered last and recommend something new based on that information. For example, when you order books

徐：所以，我觉得这是在满足一种个性化的需求。为了实现这种个性化建筑的新功能需求，数字技术可能是最有效的技术手段。我想这也是世界各个领先的建筑院校都将数字技术纳入到建筑教育中的原因。

NL：同时我也认为或许基于多方面的合作，我们作为建筑师的职责也会有所改变。例如，回到《少数派报告》，其电影设计师是 Alex McDowell，他于南加州大学电影学院的交互媒体系任职，现在与 Greg Lynn 以及来自于 Salk Institute 的神经科学家 Sergei Gepshtein 合作，致力于一种能够与我们思维互动的新型建筑的研发。这种建筑并不是一个固定的静态物体，而更多的是一个能够提供反馈的环境。我认为基于这种多专业的合作模式，建筑的定义将被革命性地颠覆。

徐：确实如此，我一直认为建筑甚至建筑教育应该与其他学科进行交流合作，因为只有我们自己多了解其他的领域，才能创造出新的知识领域。事实上，在清华我所教学的设计课程以及我的设计工作室，与多个学科进行合作研究，例如材料、结构、数控加工等。我们正研究一些新型的建筑材料，例如 FRP，同时会基于研究利用这些材料的特性建造新型的建筑，这展现了多学科合作的好处。

NL：我觉得最重要的问题是建筑师如何设计或者翻新建筑行业本身而非仅仅是建筑。换言之，世界在不停地发展与变化，为了生存，建筑师需要发明新的方式以适应新的环境。例如，在目前的经济模型之下建筑师几乎难以谋生，因此我们需要寻找新的合作以及运营方式。事实上，我想我们的建筑教育可以更加国际化，建筑教育教导我们对于空间的三维认知、材质的特性，基于这种基础教育我们可以适应许多不同的职业：工业设计师、时装设计师等等。建筑师可以是非常具有前瞻性的，但是绝大多数时候他们比较保守，我们需要抛弃过去古板的模型并以创造性的眼光看待建筑，思考建筑教育到底是什么，以及我们对于环境的设计以及与其他专业的合作中我们到底需要做什么。就目前而言似乎唯一适者生存的方式是类似于 MIT 媒体实验室中运用的那种多学科的交互合作模式。

徐：让我们回到数字建筑教育这一问题。数字技术一方面在影响着传统的设计，它跟传统的设计方法和设计思想正在结合，它在蔓延

through Amazon they will make recommendations like 'Those who bought this book also bought this book.' There is therefore a totally different structure of operations for shopping. Actually, there is a movie, Minority Report, directed by Steven Spielberg that came out in 2002. It's a very intelligent movie because it is not at all a science fiction movie but one that attempts to predict the future in an informed way. Spielberg involved scientists from MIT Media Lab and other informed experts to predict what we would be likely to find in the future. Indeed some of the things were predicted in the movie have been proved accurate. For example, the film depicts a device that will respond to your hand movements, in a way that is exactly comparable to Kinect. The film predicted that we could recognize people by their irises. Well, that's what we do now when we go through immigration control. And likewise we can now go into shops that recognize us. So this new technological world is revolutionizing every aspect of our lives.

XWG: In fact, this is appealing to personal demands and individual requirements. In this sense, digital technology is the best way to reflect and respond to tailored demands. I believe that's another reason why we should include digital technology in our education system in the leading architectural schools in the world.

NL: I also think maybe what we do as architects will change, because there will be other collaborations happening. For example - to go back to Minority Report - the designer for that movie was Alex McDowell. Alex McDowell is part of Interactive Media Department in the School of Cinematic Arts at USC. He is now collaborating with Greg Lynn, and Sergei Gepshtein, a neuroscientist from the Salk Institute, and they are trying to develop an architecture that will respond to what we are thinking. What we are talking about now is not designing buildings as fixed, static objects, but rather responsive environments. We are designing the behaviors as much as we are designing the physical object. I think there will be new forms of collaboration between multiple disciplines. What we call architecture will go through a fundamental change.

XWG: When we are talking about the discipline of architecture, we have the feeling that architecture should cooperate with many different disciplines. If we are only advancing by ourselves, we will be missing a lot of ways of exploring new boundaries. In my studio, we are cooperating with civil engineers, material experts, and digital fabrication factories, searching out new types of materiality and methods of construction. We've already built a building based on this research using FRP. This project is a good example of how to demonstrate the potential of cooperation between different fields.

并正与已有的东西交融，推动已有的建筑设计发展；另一方面，数字技术又在创造自己新的数字建筑，这个表现为：因为有了新的数字技术，一种崭新的不同于传统的建筑正在涌现。这反映了数字建筑发展的两个特点。一是蔓延，与现有的建筑的结合；另一个是创造一种新的建筑。请问，你对数字技术与建筑设计的结合有何看法？

NL: 我觉得这里存在许多不同的可能性而非唯一的结果。这也是为何我对Patrick Schumacher所谓的全球性的“参数化主义”持怀疑态度。当然我也不认为会像他所描述的那样有一个全球性的风格，建筑永远有多重的表现方式，如果建筑趋同，或者我们居住在一个完全由盖里和福斯特设计的城市中，这或许过于夸张，世界中有多种类型的经济体，不同的环境系统等等，建筑则会对其产生不同的反馈和表现方式。目前，我们能看到许多混合的建筑手段，例如：袁烽目前在试验如何利用高级算法来实现新型的建筑应对方式。在中国，由于国情的原因可以观察到高科技的数字设计方式以及传统现实的建造方式之间的拉锯。但是，事物是在进化与变化中的，就个人而言，我对于这个技术对于我们是思维模式、思考方式，还是建造方式的影响以及能走多远非常感兴趣。

徐：你的意思是在数字技术下，设计依然会存在地域性？当设计牵扯到地域问题的时候通常我们会比较谨慎，请问，你是否认为数字建筑依然还会存在地域特点？

NL: 首先，我对于地域主义的辩证性存怀疑态度，因为它们常常以风格的方式表现。当我们以行为以及过程来作为评价标准时，我们有不同的建筑应对方式，而非不同的建筑风格，他们在不同环境中有着不同的行为应对方式。我们关注的焦点应当从风格转到对于行为模式的讨论。这种思考方式将会改变建筑。

徐：我们并非只局限在建筑的形体设计上，同时，我们也考虑到文化及需求的问题。

NL: 文化理论学家并非仅仅讨论物体本身，他们关注的是我们如何在主观上与物体互动。在主观上，通过我们对于物体的观察而改变我们对其的理解，但是只有极少数的建筑历史理论学家真正将其纳入考量。除了仅仅关注建筑物体本身，关于地域主义有不少其他的

NL: I think the important issue now is how architects design or redesign not only buildings, but also their own profession. In another words things are changing rapidly. In order to survive, architects need to develop new ways to adapt to new conditions. For example, the present economic model that architects are following is almost impossible to sustain. There certainly seems to be need for new form of collaboration and new forms of operation. I'm actually thinking that architectural education could become more universal. I think architectural education teaches you about 3 dimensional awareness of space, materials and, based on that background education you could become anything from an architect, to a product designer, a fashion designer, or whatever. Architects are supposed to be very forward thinking, but most often they are very conservative. I think we need abandon the obsolete models of the past and think very inventively about what is architecture, what is architecture education and what do we do, both in terms of the kind of environments we design and the kind of collaborations we are undertaking with other disciplines. It seems that the only way to survive is to adopt some form of for interdisciplinary collaboration of a kind much like what they are doing at MIT Media Lab.

XWG: Let's go back to the role of digital technology in architectural education. Because I find that on the one hand digital technology has been influencing the traditional method of design and education, and propelling the development of existing approaches to design. And, on the other hand, digital technology has been developing a new type of architecture. In other words, as a result of developments in digital technology, there is an emergence of a new form of architecture that is fundamentally different from before. Thus, this reflects the impact of digital technology that on the one hand is spreading and infiltrating the current design method, and on the other hand generating new possibilities for architecture.

NL: I think we are open to many possibilities. It's wrong to think there is only going to be one response. That's why I'm skeptical about Patrick Schumacher's notion that there is going to be a new universal 'style' called Parametricism. I think he's wrong about his understanding of the term, 'parametric', anyway, and he's wrong about the idea of there being a universal style. Architecture will always have many different manifestations. If architecture is all the same, or if we were to live in a city where all buildings are designed by Frank Gehry or Norman Foster, for example, it would be too much. There are different economies in the world, different environmental conditions and so on. There will be many manifestations of that. At the moment we see some hybrid form of construction. For example, Yuan Feng is experimenting in how we

讨论，同样的还有针对不同的地域环境设计不同的建筑的必要性。这些争论是过时的，因为他们仅仅关注于建筑的外形而非人们如何主观上使用建筑空间的。设想一下，一个同样的高层体块，在中国、美国，以及东欧，基于人们不同的认知和使用功能模式可能成为完全不同的建筑。

徐：当你谈到人们的行为，通过对行为的分析来设计建筑，我们是不是可以这么认为，不同的地区的人在不同的文化的熏陶下所表现出来的行为是不一样的，于是所设计的建筑也是不一样的？

NL: 我完全赞同，但是区分度应当基于过程与行为而非单纯的外形与风格，就像地域主义或者参数化主义一样。

徐：让我们再讨论一下世界各地的建筑院校。2004 年以来我们一起共策划了四届北京双年展，今年是第五届。纵观过去十年全球建筑院校的发展，我们发现有些院校逐渐退出了人们的视线；同时，有些新的院校进入了人们的视野，那么你认为究竟怎样的学校才可以称得上是一所先锋学校呢？

NL: 我个人更倾向于讨论学校本身而非学校类别。我注意到，起源于 AA 的新型合作模式已经在全球蔓延开。屈米将其带到纽约的哥伦比亚大学，其他人也复制这个模式。结果就是我们得到了一种新的蔓延开的学校类型。当然在其中任教的老师也在全球遍布。我在许多不同的学校里面任教：AA、南加州大学、包豪斯、IaaC, 以及同济大学，其背后自成一种网络。很多时候，我在不同的学校之中遇到相同的人，因为他们也是这个网络的一部分。因此，网络模式是一个关键，不同学校之间也存在不同的运转速率。 有一些学校对于新的想法接纳速率以及适应速度更快，也有一些学校以不同的速率发展变化着。

还有，我注意到现在的教育市场有一些有趣的转变。当我们讨论中国学生的时候，我们需要了解中国学生在做什么，在哪里学习，以及他们会将什么带回中国。因为中国学生在西方教育之中扮演着越来越重要的角色，去年在南加州大学，我在非职业教育项目中带了 32 人的小组，而其中 27 人是中国学生。目前的事实是中国学生在逐渐改变西方的教育模式，这归功于中国经济的发展与中国人消费

might develop different adaptive methods to advance computational design. So in China, for example, there has to be some negotiation between high-tech forms of computational design and the straightforward pragmatic world of building construction. But things will evolve over time and are about to change. Personally, I'm interested in what is the far reaching aspect of this technology, in terms of how we think, how we design and how we construct.

XWG: Do you mean that as regards the digital technology part of architecture, there is still a regional difference. When we speak of regionalism we must be very careful. Do you think that digital design should also have their regional character?

NL: First of all, I'm very skeptical about the debates about critical regionalism, which often seem to me about styles of architecture. When we think buildings in terms of behaviors and processes, we again get different behaviors in buildings, not different styles of building. They are going to behave differently in different conditions. We need to shift the debate away from styles or representations towards behaviors. I think that's what will change for architecture.

XWG: Let's move on and talk about cultural issues.

NL: Cultural theorists talk not only about objects, but also about the subjective way in which we engage with those objects. The subjective lens, then, by which we perceive an object can therefore radically change our understanding of that object. But few architectural historians and theorists take this into account. Instead they focus just on the architectural object. Many of the old arguments about Critical Regionalism - and the need to have different buildings for different places - are therefore obsolete because they focus only on the form of buildings. They fail to take account of the subjective way that we inhabit buildings. You can have the same building - let's say a tower block - in China, the United States and Eastern Europe and it become very different building depending on how we use and perceive it. As we use buildings differently they effectively become different buildings.

XWG: When you talk about behavior, and say that based on the behavior we'll create architecture accordingly. Does that mean that there are different behaviors for people based on cultural differences in different parts of the world that result in a diversity of architectural designs?

NL: Obviously. I totally agree. But that diversity should be based on processes and behaviors, and not on mere styles or '-isms', as in Critical Regionalism or Parametricism.

XWG: So, let's talk about schools of architecture around the world. Since 2004, we have curated four exhibitions for the Architecture Biennial Beijing. Our fifth exhibition is this year.

能力的提升，以及相对的西方经济在某种程度上的衰退。另一方面，这也源于中国的独生子女政策，导致每一个家庭都希望尽可能多地投入到儿女的教育中。这极大地改变了西方教育，中国学生在近几年中也变得越发的成熟。

我去年秋天参加了 Ben van Berkel 在哈佛大学 GSD 的期末评图。他的小组中，最好的两个学生都来自中国。很明显，中国学生逐渐成为西方建筑学校的主力，这种教育旅游在逐渐发展，更有趣的是，让我们想象一下这些学生回到中国后会发生什么？当然，短期来看，西方院校借助大量的留学生以及高昂的学费盈利，但是最重要的是长期的影响，即当这些学生回到中国后能带来什么。例如徐丰，他在回国前曾就读于 AA DRL，之后在扎哈·哈迪德事务所工作，他将这些所学的知识带回了中国并且极大地影响了中国设计，他创建了第一期的参数化研习班，并且带来新的文化在中国迅速蔓延。因此，我认为在中国，有一种新型学校为这种新的市场培育人才并且影响中国建筑文化，最终将带来巨大的变革。

因此，很多时候哈佛大学、AA、SCI-ARC，或者南加州大学并不重要，在其中循环的是相同的理念，而且很多时候甚至是同样的一批老师在任教，但最重要的是当这些理念带回到中国，他们将从根本上改变建筑产业以及教育。目前而言，最为成功的是那些可以从根本上适应新的历史环境的。就我个人而言，我觉得最有意思的项目是，斯图加特大学 Achim Menges 所带的 ICD 项目，与德国其他的项目（例如 DIA）相同，这个项目完全免费。但是硬件资源丰富并且拥有世界上最好师资资源。当然，也有其他的地方例如哈佛大学受益于本身的品牌效应；类似的还有 MIT 的媒体实验室或者 AA。这成为了一个全球性的举措。AA 是一个很好的范例，AA 基于其在全球数量繁多的访问学校，使其成为一种文化而非学校本身。因此，我们不应该关注于特定的某一个学校而是应当将目光投射于一个拥有巨大效应的全球性的浪潮，尤其在中国。

徐：事实上，有的学校对数字技术是接受的，有些学校是反对的。是否这种对数字技术的接纳程度可以在某种程度上代表着这个学校发展的未来呢？你又是如何看待对于数字技术的这两种不同的态

Which schools are the leading schools of architecture in the world based on our curatorial experiences over the past ten years? In fact, some architecture schools have disappeared from our selections while other new ones have been added. And so what's your opinion about the leading schools of architecture in the world?

NL: I would prefer not to talk so much about actual schools but rather types of school. It seems to me the progressive networked model that was first developed in the AA has begun to spread. Bernard Tschumi took it to Columbia in New York, and others have copied it. What we are getting is a type of school that's spreading around the world and of course the people that are teaching are spreading around the world. I teach in many different schools, in USC, Bauhaus, IaaC and Tongji. There is a kind of network. Often I came across similar people in different schools because they are also a part of that network. So the network operation is the key thing. Maybe also there are different speeds at which the different schools are operating. Some schools recognize new ideas and quickly change and adapt. The schools on our list are those that have adapted relatively quickly, but there are other ones that are maybe operating at a different speed.

However, I also think that an interesting market has developed for education. When we are talking specifically about Chinese students, we have to look at what Chinese students are doing, where they are going, and what they are bringing back to China, because Chinese students are becoming the most significant players in Western education. At USC last year I was teaching 32 students in our post professional master's program, and 27 of them were Chinese. What's happening is that Chinese students are changing the nature of Western education. This is partly because they have increase spending capacity as the Chinese economy is growing stronger, while the Western economy is in some form of recession. But it is also partly because here in China you have one child per family, which means every family wants to invest as much as possible in the education of their one son or daughter. That is really changing the way that things are happening. Plus of course Chinese students are getting much more and more sophisticated.

I attended a review of Ben van Berkel's studio at Harvard GSD last autumn, where the two top students were Chinese. It is clear that Chinese students are beginning to dominate many Western schools of architecture. There is a form of educational tourism developing, and what I find interesting is what happens when these students come back to China. Of course, in the short term Western schools are benefiting from the fees being paid and the number of students, but the most important factor is the long term question of what these students do when they come

度？你认为反对数字技术的传统院校的未来是否依然光明？

NL: 我曾经在剑桥大学建筑学院学习，在那里，我们的设计课不允许使用计算机来辅助设计，老师还教导我们这种技术使人们疏离。由此，整整一代的建筑学生几乎找不到工作，因为他们不会那些在办公室里面必须的软件。对我而言很明显我们不能忽略这些技术，就如同我们不能忽略手机以及我们不能忽略网络给我们带来的改变。我们不能抗拒历史的潮流，它在多种层面上影响着我们的生活，因此，我对于有些对于技术持有负面态度的教授感到担忧。他们认为这些技术很肤浅，并且不考虑当地的政治以及社会问题等等，他们甚至觉得这不是设计。我认为这是一种老套的见解，世界已然完全改变，许多这类教授就像是从恐龙时代来的老古董，而学生则是他们言论的受害者。我认为我们有责任为学生提供适应当今社会的教育年使他们了解新的技术发展趋势。同时，我也认识到我们需要客观地评价这些技术，我们不应当仅仅因为事物是新的而重视，我们应该以批判性思维的方式评价它们以及与其互动。因为我也承认技术的使用具有一定的负面作用，但是同时我们也需要以一个开放性的姿态来面对它们所能给我们带来的可能性。因此，结论是我们应当置身于两个阵营之间，即完全被新技术吸引但又无法批判性地看待它们，与对其完全拒绝这两种态度之间。

徐：是的，我同意你的说法。让我们重新回到今天的主题——设计智能：高级计算性设计研究。这里面重要的是高级计算性，举个例子，机器人的应用。如今，越来越多的学校和机构开始对机器人的使用进行研究，用高科技的手段来加工建筑构件并实现建筑结构。你如何看待机器人技术这项新的研究趋势？

NL：事实上，我个人也是美国航天局在南加州大学的机器人项目组的成员。我们的目标是研发一个可以在月亮或者火星上三维打印结构的机器人，由此，我对于您提到的那个领域有亲身的接触。但是我认为许多学校对于机器人有着错误的期望值，有几个学校简单地买了机器人之后就期待它能带来直接的变革，但是就机器人而言，我们需要克服以下几个难题。第一，需要购买机器人；第二，我们需要对其编程；第三，机器人需要一个严格控制的工作

back to China. For example Xu Feng, who studied in AA DRL, and then worked for ZHA, came back to China, bringing with him that knowledge that had a radical impact on Chinese design. As we know, he started the first digital workshop, and initiated a culture that began to spread like a virus throughout China. So I think what is happening is that there is a new type of school that is catering to this new type of market that is feeding back into Chinese architectural culture, and radically changing things. So sometimes it doesn't matter if it's Harvard, AA, SCI - Arc, USC or whatever. It's the same idea that is circulating, and it's the same professors in many ways. But what is happening is that these ideas are coming back to China, and completely revolutionizing practice as well as education. But it seems to me the most successful ones are the ones that are really adapting to the new world. Personally, I think that the most interesting program right now is maybe at the ICD program that Achim Menges runs in the University of Stuttgart. As with other programs in Germany - such as the Dessau Institute of Architecture [DIA] - there are almost no fees. But they have considerable resources in terms of equipment and they have one of the most interesting professors in the world working there. Of course, there are other places such as Harvard that benefits from its brand name, or MIT Media Lab, and likewise the AA. So it's becoming a global operation. And maybe the AA is a good example because the AA has so many visiting schools around the world that it has become a culture more than a school. So we shouldn't identify one particular school. Rather we should identify an emerging trend that is having a global impact, especially here in China.

XWG: In fact, some schools accept digital technology, while some others are against it. Does the application of digital technology, under current scenario, indicate the successfulness of the schools in the future? How do we evaluate this attitude toward digital technology? It there a future for the traditional school that rejects digital technology?

NL: I studied at the University of Cambridge School of Architecture. We were not allowed to use computers in design studio. We were told the technology was alienating. And so - as a result - a whole generation of students graduated that was almost unemployable, because they couldn't use the software needed in offices. It's very clear to me that we simply cannot overlook these technologies in the same way that we cannot overlook mobile phones or overlook the internet. They are part of the way that we operate today. We simply cannot resist the tide of history. The tide is changing and this affects everything we do. So I worry about some professors who have very negative attitudes toward these technologies. They think it's superficial, that it doesn't engage with social and political issues, and so on. They think it doesn't even address design itself.

环境，如果没有一个严格控制的工作环境我们就无法利用其做喷色等基本工作。通常，最贵的并不是机器人本身，而是建造一个可控环境以及编程。因而，我们往往面对的是一个学校常常买一个机器人作为宣传的卖点，至少在某些学校是这样的，他们利用机器人来宣传推广学校，这是公共关系的尝试，但是就目前为止，他们并没有真正非常有效地使用机器人。同时就学生而言，机器人的学习周期非常长，以至于他们停止使用机器人。同时，我认为机器人的未来可能有一定的改变，就像我们在手机的发展历程中观察到的一样，它将向更直观的交互方式转变。同样的变化我们也可以在 Grasshopper 中看到 ,Grasshopper 本质上来讲是一个基于代码的算法设计，但其利用符号使其可以直观地使用。所以我认为在未来，机器人使用起来将更为简单，就如同利用 IPHONE 或者使用 Grasshopper, 最终机器人则会由此在建筑院校之中流行起来。但是，就目前看来，机器人的使用依然很困难，它们的主要功能是为学校品牌的宣传做贡献。

徐：我认为需要有更多的人投入到对机器人的研究。事实上，对于机器人的研究仅仅是处于萌芽阶段，几乎还没有人清楚如何将机器人用到建筑建造的过程中。大家还只是在一个炫技的阶段，仅仅是做一些初级的研究。机械臂的使用最重要的是机械臂最前端的部件，机械臂能做什么取决于建筑师对这个前端部件的开发。实际上，机械臂已被广泛使用在其他领域的产品建造方面，但是，其他领域并不知道建筑师想做什么，如果我们与其他学科进行合作，这些技术便可以移植到建筑领域中来。我觉得建筑师最重要的事情是根据自身的需求来探索机械臂前端部件的使用，但是，现在许多建筑师却并没有重点关注这个问题。

NL：在这方面进行着一些有趣的研究工作。我们必须区分一般的应用与例如在密歇根州立大学的工作，其系主任 Monoca Ponce de Leon 真正了解机器人。同样的在 ETH 中 Gramazio 和 Kohler 也在从事重要研究。现在机器人的数量越来越多，同时对其编程的经验也在积累。这一方面会通过编程的普及而发展。我坚信就算现在机器人没有找到一个合适的使用方法，在未来也可以找到。但是就我个人经验而

My problem with that is this is old fashioned thinking. The world has completely changed. Many of these professors are frankly dinosaurs. They are obsolete. It's the students who are suffering, when these professors are saying these things. I think that we have an obligation to the students to provide them with an education that's appropriate to today's society, and to be open to new technologies. Having said that, I think it's important that we remain critical about them. We shouldn't welcome them just because they are new; we should evaluate them and engage them in a critical way. Because I think there are some shortcomings about the use of computation. But we have to be open-minded about the possibilities at the same time. So the answer is somewhere between the two camps, between those who completely subscribe to these new technologies but who are not critical enough about them, and those simply who deny their relevance.

XWG: I agree. Back to the main title: Design Intelligent Advanced Computational Research. The important words are 'Advanced Computational Research', by which we refer, for example, to robotic research. People from different schools and organizations want to apply this technology to the fabrication or even the construction. What do you think of this new advancing robotic technology?

NL: Well, I'm actually involved in a project in USC working for NASA to develop a robot to 3D print structures on the Moon and Mars, and so I have some direct experience with that field. But I would say that a lot of schools have false expectations about robots. There are several schools that simply buy robots and expect there to be an immediate revolution as a result. But there are three important issues to take into account when dealing with robots. Firstly, you have to purchase the robot, and then, secondly, you have to program the robot - and therefore you have to have someone who knows how to program the robot - and, thirdly, you have to have an environmentally controlled space in which that robot can operate. You can't even spray paint and do other operations with a robot without a carefully controlled environment. The most expensive item is often not the robot, but the cost of constructing the right environment for it and programming it. So what often happens is that a school will buy a robot almost like a publicity stunt. Certainly in some schools they use their robots to promote the school. It's a PR exercise, but so far they haven't really used those robots very effectively. And often the students themselves find that it takes so long to learn how to use the robots that they stop using them. But I suspect the future of robots will change slightly, in the way we can see with our mobile phones, where everything is shifting toward user - friendly operations. And the same goes for Grasshopper. Grasshopper is actually a form of

言，当前的建筑建造行业并没有像汽车制造业一样完全准备好迎接机器人的到来。在汽车生产中，每一台汽车均由生产线以及机器人自动组装而成。其中，部分的原因在于资本投资，制造业没有那么多的钱来投入某个特定型号的研发。因为每一个建筑均是不同的，同时人们的建造设计时间非常紧促，而使用周期也非常短。或许建造业就目前而言并未准备好迎接机器人的到来，但是很明显，这一天终究会来临的。

如果我们要归纳一下数字建筑的影响，最本质的是控制以及精度。建筑本身即是关于控制与精度。它包括设计的控制，例如对于曲线的控制，以及对于设计的形体加工生产的控制。在 2004 年我们找了大量的中国工人来切割加工一个数字化设计产品，原本它应当由 CNC 加工的。同时，对于运输以及生产流程的控制也是非常重要的。因而 BIM, 建筑信息模型即是用来协调各个部分，控制运输以及建造的。它完全是多重层面的控制，而我们也要控制这些技术给我们带来了什么。没有它们，未来是无法想象的，就像没有网络及手机的未来是无法想象的。

徐：实际上，我认为用机器人应做一般的 CNC 机床以及人所不能做的事情。数控机床可以做人所不能做的事情，而机器人应该能做数控机床所不能做的事情，这样才能充分发挥它的作用。但是现在许多机器人的研究是在让机器人做数控机床就能完成的事情，我觉得这样的工作有点浪费。

NL：我认为最重要的是认识到我们的目标是什么。例如当我们使用谷歌搜索的时候，原本需要几个月甚至几年去收集的信息可以飞快地呈现在我们面前，有一些事情只有计算机才能完成。事实上我认为计算机很呆板，机器人本身也很呆板。我们需要对其编程，使得它们能够以一种聪明的方式来完成人们所无法做的工作，或者以更为有效和优化的方式完成任务。在这方面我完全同意您的意见。但是我同时也认为我们现在还处于计算机发展的起步阶段。我最为欣赏的对于计算机的评价是 Ben Bratton 所说的我们还在“计算机技术的默片时代”，这即是我们的进程。

徐：很有意思的观点，事实上，计算机刚刚处于发展的萌芽时期。

algorithmic design that relies on code but uses visual icons to make that code accessible. So I suspect that in the future, robots will be much easier to use. It will be like using an iPhone or working with Grasshopper. And robots will therefore become more popular in schools of architecture because of that. But - at the moment - robots are still very difficult to use, their primary purpose seems to be to promote or brand a school.

XWG: So do you think there is a need to dedicate more people to research into the use of robots? In fact, here, robotic research has just begun. No one knows, for example, how to use the robot for architectural fabrication. We can only use it to show off, or doing primary research with it. I think the most important thing in the field of robotic research is not the robot itself, but rather the head of the robotic arm. What do we do with it is up to the exploration of the architect. There are many applications of the robotic arm in fabrication in other fields. And so, if we could cooperate with other disciplines, this technology could have a direct impact on architecture. However, for other disciplines, they have no idea what the architect aims to do. I believe it's most important for the architect to discover how to use the head to meet the architects' needs. However, at the moment, the focuses of many architects are away from it.

NL: I think there are some people doing important and interesting work. We have to distinguish, for example, the work that's going on in the University of Michigan, where Monica Ponce de Leon - who really understands robots - is the new dean, and also in ETH Zurich, where Gramazio and Kohler are doing important work, from other schools. So there is a growing number of robots, and a growing body of knowledge as to how to program them. That aspect will develop and spread widely just as the use of scripting. I'm quite convinced that even though they are not used quite properly right now, they will be in the future. But based on my own experience, at the moment the whole field of building construction is not really ready for robotic fabrication in the way the car industry is. Every car is made robotically through the automated assembly line. The reason is partly because of lack of capital investment, in that the construction industry doesn't have the kind of money to invest in particular models, and partly because every building is different and therefore everybody is operating in a very short time scale with short term operations. Maybe the condition in the construction industry is not yet ready for the full potential for robotic fabrication but it is clear that it must come eventually.

If I were to summarize the impact of computational architecture, ultimately it's about control and precision. Architecture is all about control and precision. It's about control over design

你认为50年之后的未来计算机技术会是怎么样的呢？

NL：让我们回到我之前关于这个十年的预言。在2000年比尔·盖茨预言2010年的时候，计算机将触及生活的所有方面，他的预言完全正确。我对于2020年的预测是，我们甚至不再使用“计算机数字化”这个单词因为它的使用范围过于广泛；在50年之后，我们所有的工作方式将完全数字化。但是我不想预测究竟是如何实现的，而仅仅是说我们在普及电邮之后已经不能再回到写信发电报的年代，由此向前发展的脚步是不能倒退的。但是我们也要了解这个发展的过程使其服务于社会，就像与电邮相伴而来的还有垃圾邮件等等，这并非都像我们设想的那么美好。因而我们必须了解为了人类的发展我们可以做什么，毕竟最终这将取决于我们自身。

徐：在1980年代的时候，中国人画建筑图还是用尺规和图板；在80年代中后期，就出现了CAD来辅助画图，这经过了很长时间；现在人们开始用3D模型来做设计，而并非基于CAD这种方法。那么，对于计算设计来讲，是不是有可能取代CAD以及3D设计这种方法，而完完全全直接是生成设计？你刚才谈到了2050年，会不会到那个时候，就完全是计算设计，而没有了现在的画图设计？

NL：我认为算法设计就是三维的。我不认为这是二维对抗三维，或者算法设计对抗三维设计。我的感觉是我们使用谷歌搜索的方式最终将成为主导，这样我们可以非常直观地完成某些操作。例如，让我们想象一下，我们可以输入：“两个卧室、北京、游泳池、朝南”，而计算机可以自动生成多种组合模式，我们则可以浏览并且决定到底要不要使用这个设计，这将是一个全新的生形方式。在2003年，我曾经使用过EIF形体，一个可以自身孕育结构的软件；每20分钟，新的组合方式被生成并在屏幕上显示，人们可以比较不同的方案并且选择自己喜欢的组合方式，由此，它成为一个全新的工作方式。我们几乎可以将它与我们现在照相的方式进行比较，在过去当使用手动相机的时候，我们需要仔细设计参数以得到一个完美的照片；现在我们则是利用相机拍摄很多相片再选取最喜欢的，相机本身已经自动对焦并且设定好了。我怀疑在设计过程中也会有一个类似的变化过程，这将不是一个老的只生成一个设计的过程，而是利用计

- let's say the manipulation of a curve - and then the control of the fabrication of the designed form. In 2004 we found ourselves with an army of Chinese laborers cutting and making a digitally designed structure which was originally designed to be CNC milled, and it ended up being a disaster. And also importantly it's also about the control of the logistics of construction. So Building Information Modeling is about the control of the coordination of the parts, the control of the logistics of construction and so on. It's all about control, and we simply need to have the control that these technologies can offer us. Without them, the future would be unthinkable. Just like the future would be unthinkable without the internet, or mobile phones.

XWG: I believe that the goal for robotic research is to use it to do what's normally not possible with CNC machines, and to use it for what human are not capable of, just as the CNC machine should be used to do what's not possible by human labor. In this way, we can fully realize the robot's potential. However, some robotic research is being applied for ordinary work that could be effectively done by hand or CNC machines. Personally, I feel this is a waste of the technology.

NL: I think it's important to recognize the tasks. For example, when we go Google, the search engine, it's astonishing how quickly we can search for information which originally takes month and years to find. There are things that computers can do that we simply can't do. Actually I think the computer is kind of dumb, and the robot is also dumb. You have to program it. And so they have to be used intelligently to perform tasks human can't do - to do things much more efficiently and so on. I completely agree with you in that respect. But I think we are at an early stage in the development of computation. My favorite comment about the use of computers is by Ben Bratton, who once said that we are still in the 'silent movies stage of computation'. That's where we are now.

XWG: So in fact, computation is at an early stage of development. What would be the future of computation like about 50 years from now?

NL: Well, let me just go back to my earlier prediction about what will happen by the end of this decade. In fact in 2000 Bill Gates made a prediction that by 2010 nothing would be untouched by computation. And he was obviously right. My own prediction is that by 2020 we won't even use the word 'computation' because it will be everywhere. In 50 years time the whole way in which we operate will be totally digitized. But I wouldn't want to predict exactly how, except to say that this is obvious in that even now we can't go back to using telegrams and writing letters once we have developed communication through emails. So it's moving forward irreversibly, but we have to understand the processes in order for them to work for the benefit

算机研究不同的可能性，并且选择理想解答的过程。

徐：看来计算设计必然会代替传统设计从而给建筑设计带来崭新的设计方式。

of the society. One of the early predictions about computation was that it would automatically lead to a much better world. In fact what happened immediately was that we got spam in our email accounts, and so on - not at all what we expected. So it's important to understand what we can do with it and make it work for the benefit of humanity. It's up to us in the end.

XWG: In the 80s, people would still make architectural drawings with pencil and ruler. At the end of 80s, CAD had emerged to assist in the drawing process. Through a long period of development, many people now use 3D modeling to design instead of using a CAD based process. For algorithmic design, do you think it's possible to replace CAD and 3D form making and have completely generative design? By 2050, do you think that all design will be algorithmically based instead of developed from drawings?

NL: I think that algorithmic design is 3D, so I don't think this is a question of 2D vs. 3D, or algorithmic design vs. 3D. My hunch is that the way that we use search engine such as Google will come to dominate, and that what we will be able to do is something very straight forward. Let's say, maybe we can input: '2 bedroom house, Beijing, swimming pool, south facing,' and the computer will be able to generate many options. We'll be able to search through them and decide whether to use this design or that one. There's probably going to be an entirely different way of generating form. Already in 2003 I had the experienced of working with EIF form, a structural software that can generate or breed structural forms. Every 20 minutes, some new configuration would be generated. It would crystallize and appear on the screen. And then you could simply look at different ones and select your favorite one. So it became a very different way of operating. We can almost compare it to the way that we take photographs these days. In the old days you had your manual camera and you would carefully set it up to get the one perfect shot. Now we take lots of shots with a digital camera and simply select your favorite one. It's already automatically focused and everything. I suspect there is going to be a similar shift in the design process. It's not going to be a question of the old fashion way of producing just one design but of using the computer to search different possibilities, and then selecting the preferred option.

XWG: For sure computation is definitely going to develop a different way to produce architectural designs compared to the traditional methods.

美国哥伦比亚大学 /Columbia GSAP

内在与外放 / Narcissus and Nemesis

这个项目探索了一种理想化的愿望：通过分析曲线中内在的力量，并在生长过程中寻找精确度，以真实地呈现并创建一个复杂的几何形体。这一形式的生成是将两种不同的“编码”组合在一起的结果。第一种“编码”定义了一个从“初始位置”到“计算后位置”的轨迹。第二种“编码”定义了一种关于轨迹的“层理法”，这基于对曲线曲率（半径和向量）或曲线法向的分析。

The project explores the idealized desire to achieve true reflection and suggests the creation of a complex geometry that occurs through the analysis of reflection as well as of the embedded forces within a curve, questioning notions of precision through growth. The form generated is the result of the combination of two different codes. The first code defines a trajectory from a starting location through calculating reflection. The form is generated from the extrusion of trajectory paths created via the first code. The second code generates a stratification of trajectories based on an analysis of curvature (radii and vector) or through an analysis of the curve's normal.

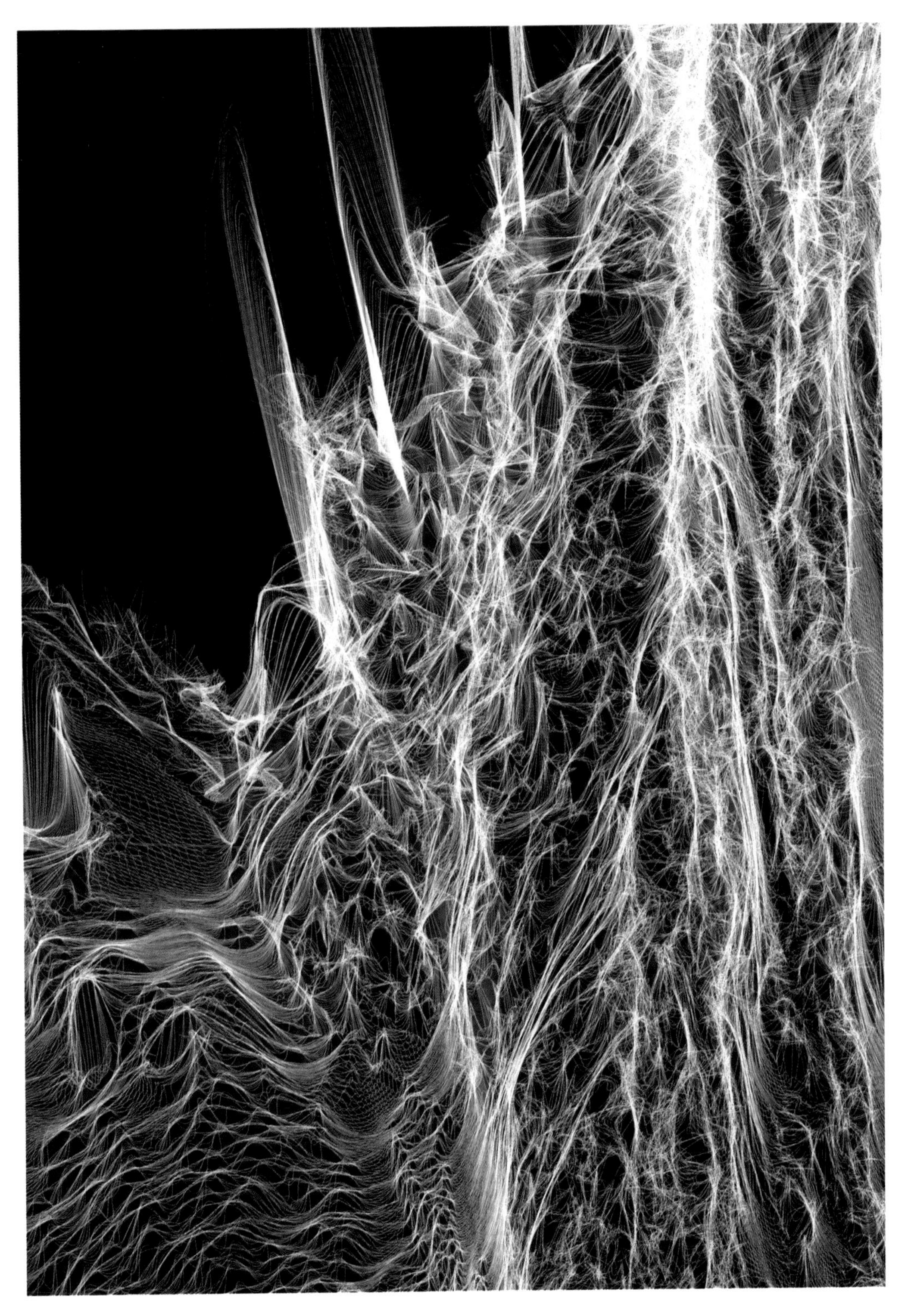

脑波砌码 / Brain Hacking

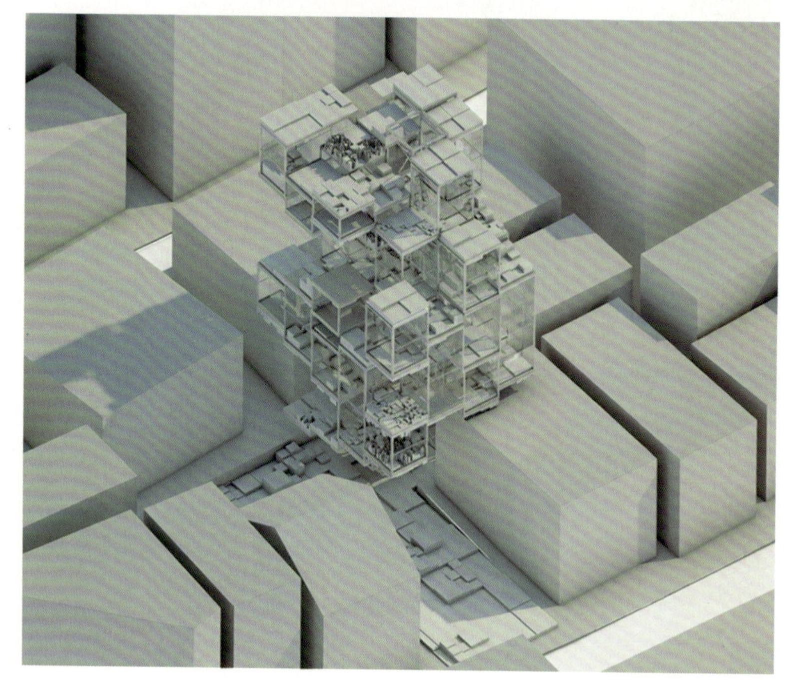

这是一个东京的城市住区，被设想成为一个信息的中枢，可以积极地应对居民们的健康、认知过程和行为。通过本工作室的“人脑 - 电脑”互动界面，把关于隐私、自我、自由意志和后人类生活方式的观点和交流编织在了一起。本工作室与 Neuromatters 公司合作，启动了一种技术，其所创建的系统利用具有闪电般处理能力的视觉皮层，去识别图像和为图像的活跃程度排序。工作室重新改变了系统的原有意图，运用 Processing 软件进行整合，形成了一个生成空间的工具。用此工具生成了大量的备选设计方案，并快速对其进行评审和仔细检查。

An urban dwelling in Tokyo is conceived as an informational appliance that will actively intervene in the health, cognitive process and behavior of the houses occupants. Themes and conversations on privacy, self-hood, free will and post-human lifestyle are knit through the studio's work with brain-computer interfaces. The studio collaborated with Neuromatters, a tech startup that is building a system utilizing the lightning-fast processing of your visual cortex to identify and rank salient characteristics of images. The studio repurposed the system into a generative space-making tool by integrating it with Processing, generating thousands of design candidates to rapidly review and scrutinize.

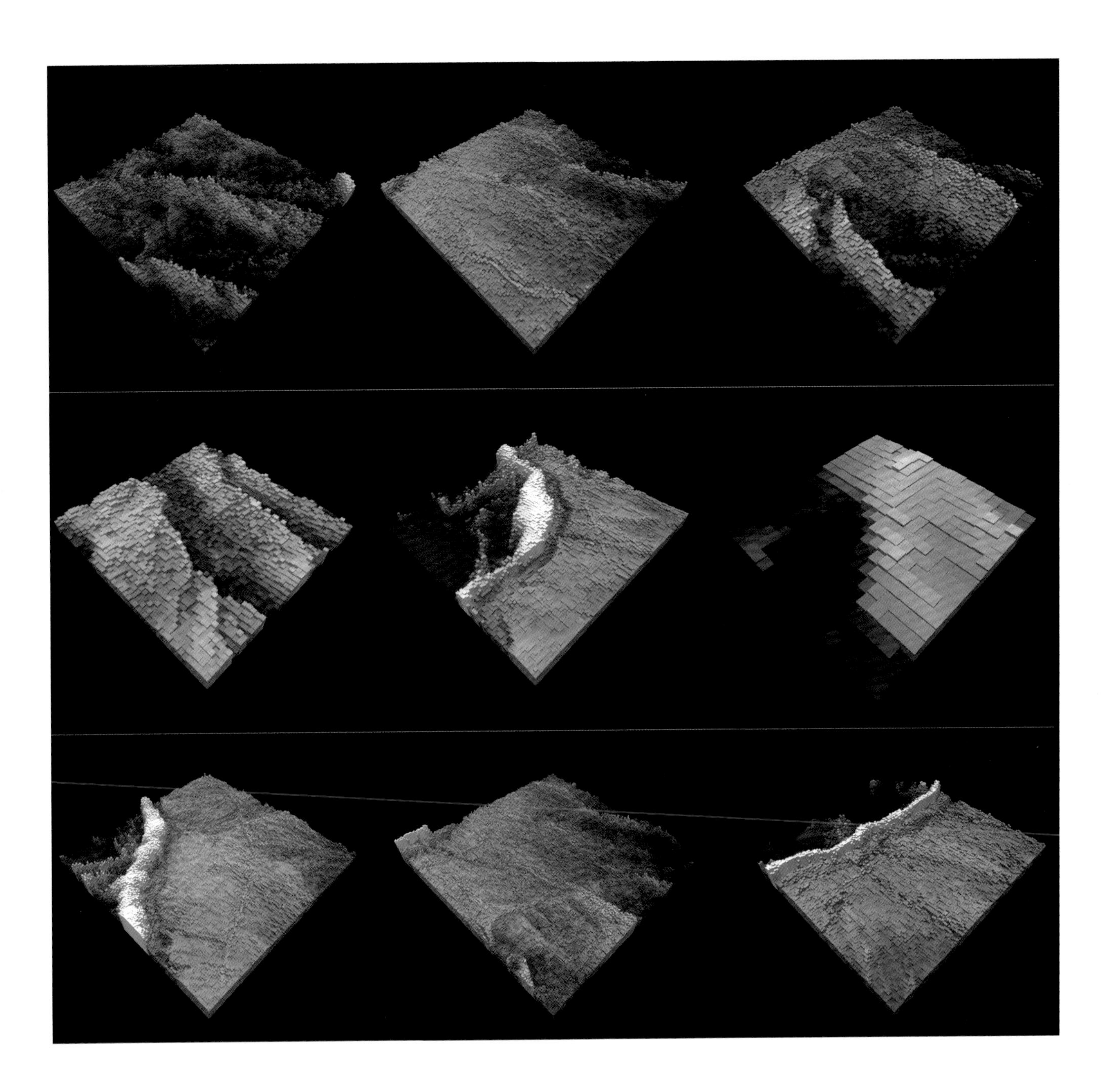

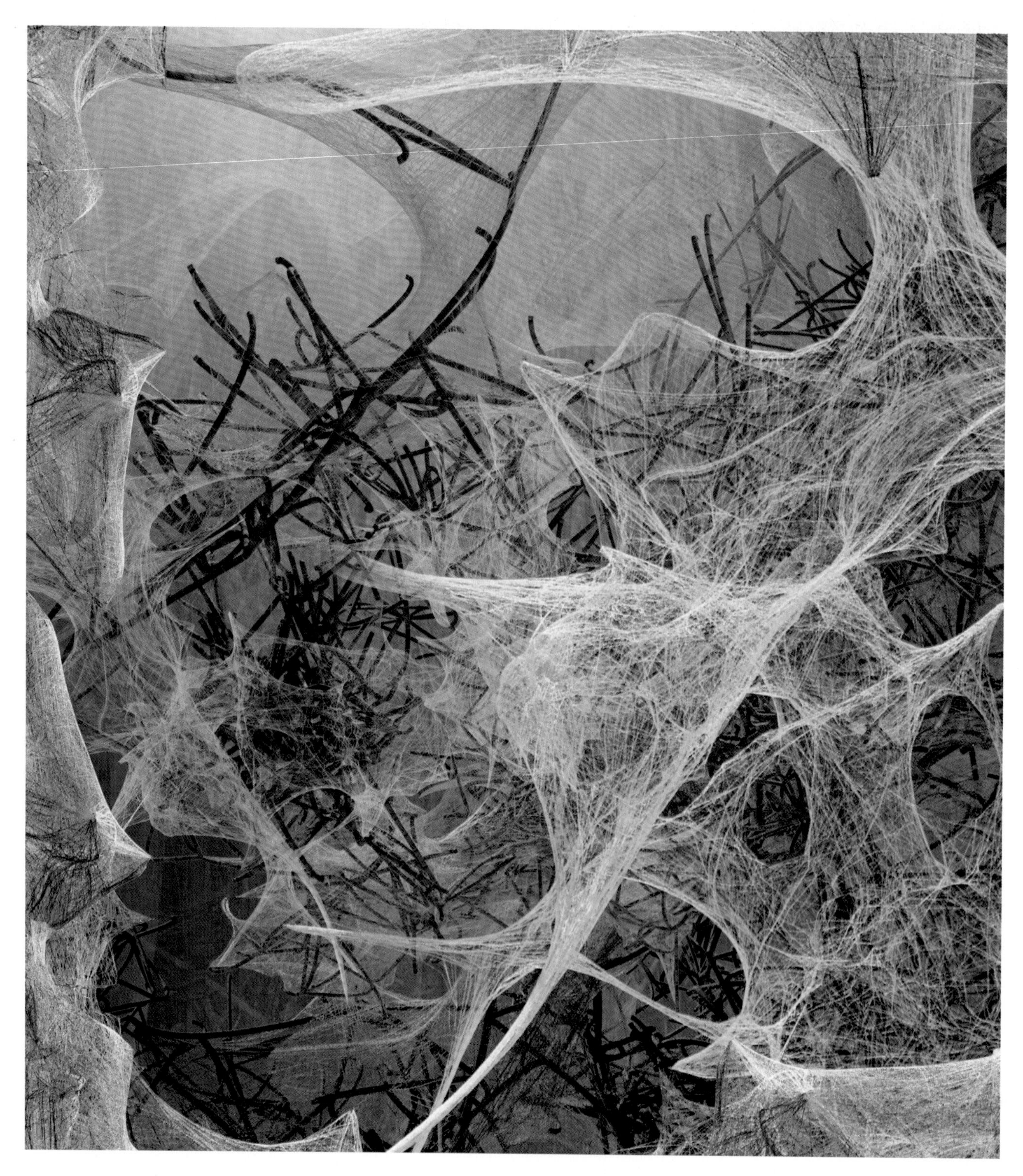

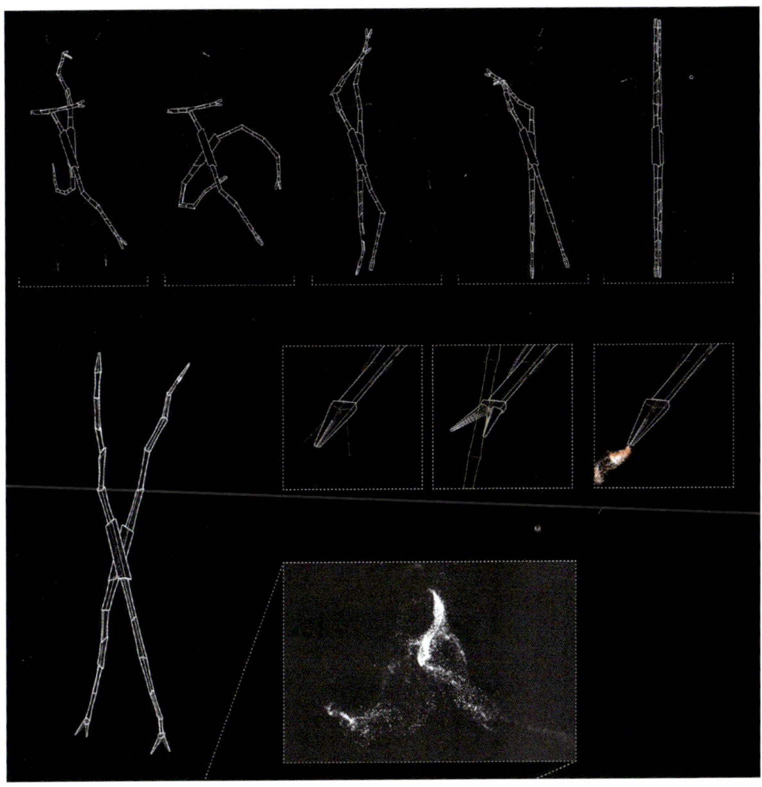

家园中不仅只有人类 / No 'Heimat' for You

这个项目利用农艺来为劳动者建造一个以甘蔗作为结构支撑的社区。这个社区不仅是人类的居住地，同时也是多物种密集的地方（包括蜘蛛）。相互协作的代理系统呈现出的“摧毁力”重新定义了空间的 DNA，以帮助取得丰收。利用植被的循环可再生，人们可以收割并取得收成，动物也同样如此。当甘蔗结构系统开始分解瘫痪，蜘蛛会由于害怕掉入水中而处于对死亡的极度恐惧中。当新一轮的甘蔗生长起来，物种会从已毁的结构中重新涌现。

This project uses agriculture with the intention of creating a structure of sugar cane for the workers, a community based not only of humans but also a high intensity of creatures including spiders. Corruption appears as a collaborative agent in order to redefine the DNA of the area, helping to harvest the agriculture. The loop of vegetation can be harvested, as well as the creatures themselves. As the sugar cane structure system begins to decompose, the spiders are driven by their fear of water, a schizophrenia of death. A new harvest of sugar cane grows under, and then re-emerges through the decomposed structure.

美国哈佛大学设计研究生院 / Harvard GSD

水平住宅 / Horizon House

当代的住所必须扩展它的热力学和生态学界限。为了理解它们代表的有活力的过程，它需要本地经济和采购制造技术部件的居民共同参与。水平住宅采用当地收获和打捞的木材，代替蕴藏能量高的材料，如：混凝土。这个项目陈述回归自然的概念，通过合并模拟分析和灯控传感器创造未来的性能。使用者的活动通过辐射和地面储存系统形成热舒适的外壳，只通过当地森林的产品燃烧提供能量。

The contemporary dwelling must expand its thermodynamic and ecological boundaries. It requires the engagement of both the local economy and the inhabitant in the sourcing and manufacture of its tectonic components, to understand the energetic processes they represent. Horizon House incorporates locally harvested and salvaged wood, replacing high embodied energy materials such as concrete.

The project addresses the concept of retreat in nature by combining simulation analysis and post occupancy sensors to manage its future performance. The activities of the user shape the thermal comfort envelope through radiant and ground storage systems, only powered by the combustion of local forest by-products.

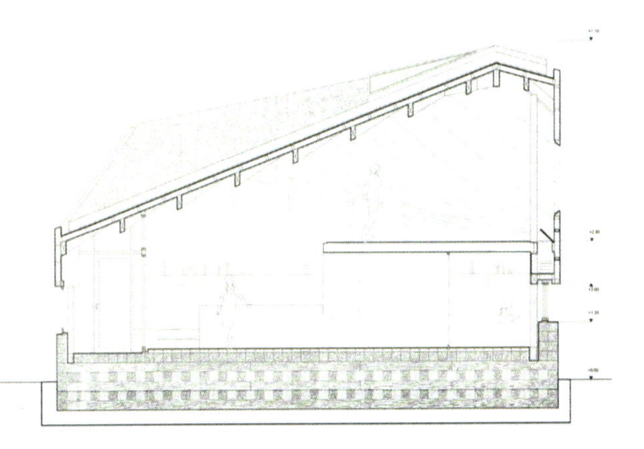

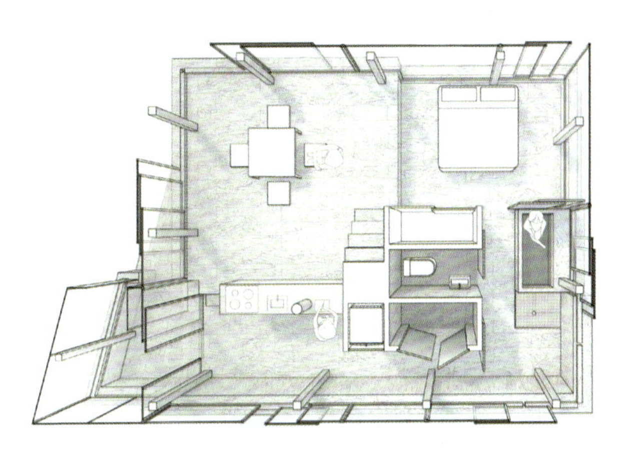

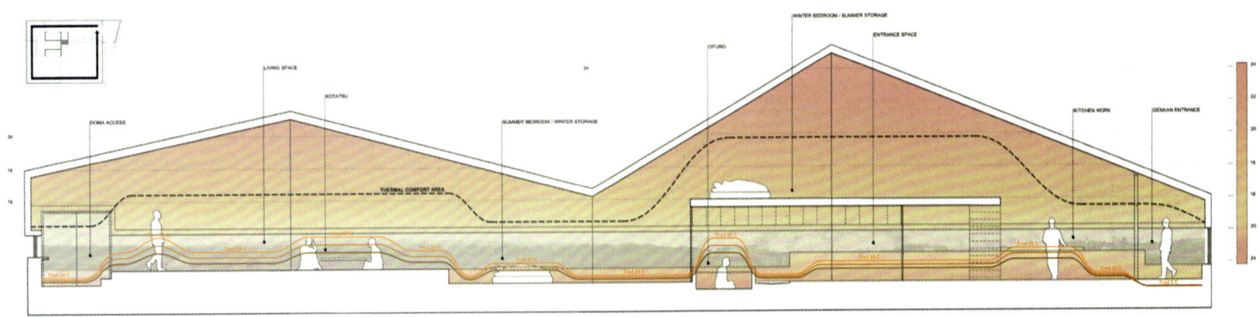
WINTER BEDROOM / SUMMER STORAGE
ENTRANCE SPACE
OFURO
LIVING SPACE
KOTATSU
KITCHEN WORK
GENKAN ENTRANCE
DOMA ACCESS
SUMMER BEDROOM / WINTER STORAGE
THERMAL COMFORT AREA

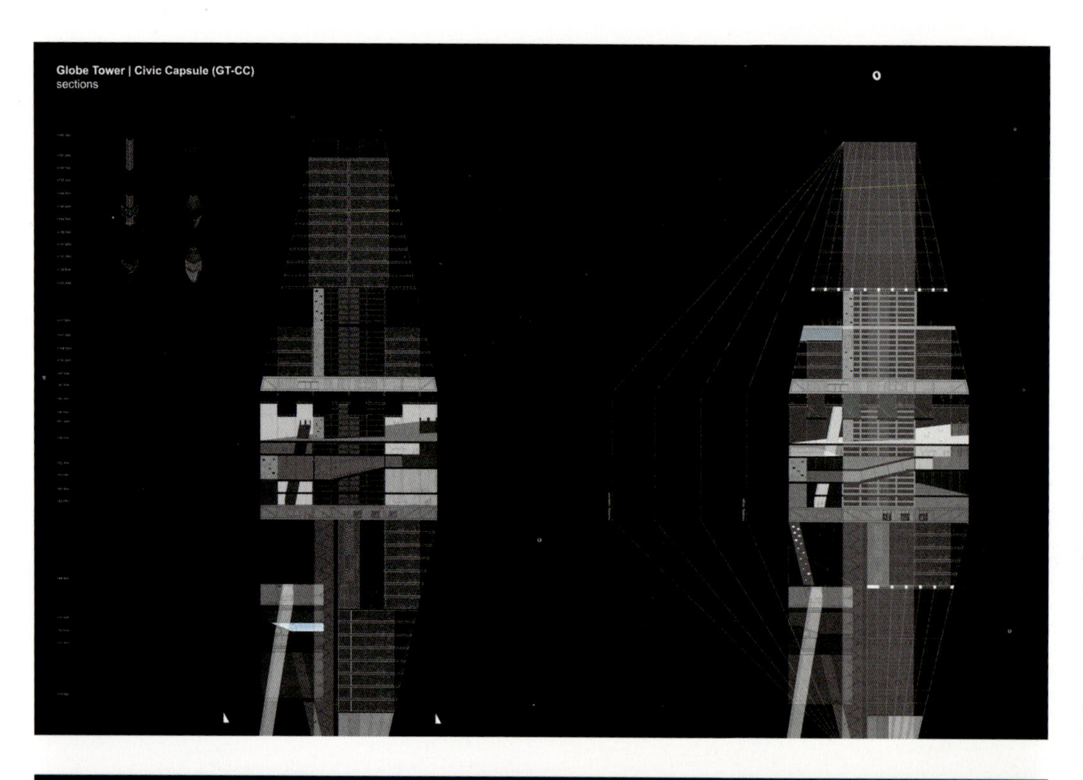
Globe Tower | Civic Capsule (GT-CC)
sections

成为另一个 / Becoming-Other

城市化是一个具有无限可能性的领域。建筑是其特异性。城市化具有持续发展的潜力。建筑使持续化成为可能。是否有两者形成过程中的交点？这个交点会是什么？这个项目涉及多个计算技术的应用。应用 Grasshopper 使城市策略系统化，并测试这个系统的多个分支。之后，在设定建筑雏形时，应用Rhino和Grasshopper，在Z-corp制作三维打印雏形。

Urbanism is all about potential: a domain of endless possibilities waiting to be actualized (to become-something). Architecture is all about specificity. Urbanism is in a constant state of becoming-potential. Architecture is in a constant state of becoming-reality. Is there a point of intersection of both becomings? What would that point be? This project involved the use of several computational techniques. For the Urban Strategy, Grasshopper was used in order to systematize and test multiple ramifications of the system; and later, during the definition of the architectural prototype, Rhino and Grasshopper also were used together with 3d printing of prototypes in Z-corp.

牧民，技师 / The Nomad, The Technologist

今天，领域的建构是由场地形状和地理的景观数据所决定的。该项目利用遥感控制系统、实时数据和模型进行一套反向工程算法计算，构建一个灵活的、游牧的城市框架。应用方程输入环境变量创建在内蒙古 133 个游牧自由贸易区的结构和布局。它通过 GIS 里的地理数据，Processing，与实时传感器相连接的 Grasshopper，生成 133 个场地的季节性编队，用来自动部署和计算预计的游牧民族的运动情况。

Contemporary constructions of territory today are made through landscapes of data that shape site and geography. This project leverages the cybernetic systems of remote sensing, real time data and modeling to reverse engineer a set of algorithms that construct a flexible, nomadic urban framework; equations that allow for variable ecological inputs to create structure and layouts for 133 nomadic free trade zones in Mongolia. It deploys the automation and computation of projected nomadic pastoralist movement through geographical data in GIS and Processing, alongside a model with real time sensor data via Grasshopper to generate seasonal formations of the 133 sites.

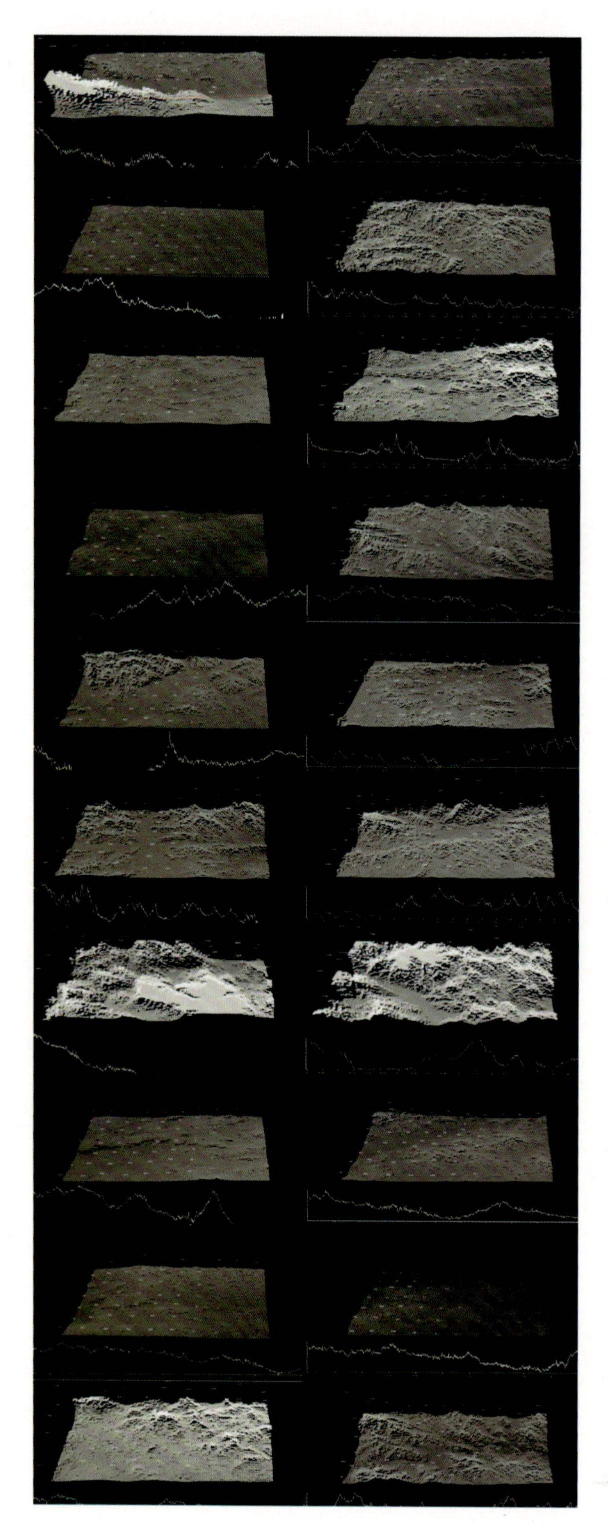

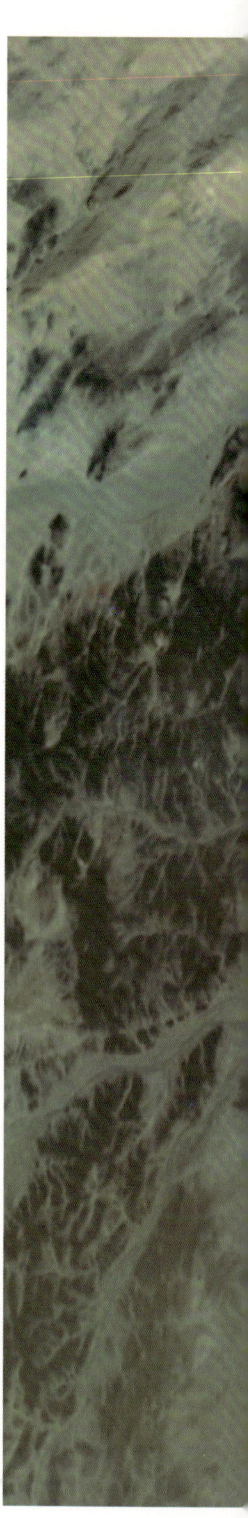

美国麻省理工学院 / MIT

绸亭 / Silk Pavilion

由 MIT 媒体实验室的 Mediated Matter 研究组开发的“绸亭”旨在探索数字技术和生物制造间的关系。先前的结构创自 26 个由丝线制作、CNC 切割的多边形镶板。受蚕的那种能够用单一多属性的丝线生成三维蚕茧的能力启发，整体介入的几何体通过使用一种算法创建，它可以将一条单一连续的线穿过不同的块，并调节其密度。一个由 6500 条蚕组成的集群被放置在脚手架的底部边缘，织造平的非缠绕丝块，同时它们增强了 CNC 预存丝绸纤维的局部间隙。

Developed by the Mediated Matter research group at the MIT Media Lab, 'The Silk Pavilion' explores the relationship between digital and biological fabrication. The primary structure was created of 26 polygonal panels made of silk threads laid down by a CNC machine. Influenced by the silkworm's ability to generate a three-dimensional cocoon out of a single multi - property silk thread, the overall geometry of the intervention was created using an algorithm that assigns a single continuous thread across patches providing various degrees of density. A swarm of 6,500 silkworms were positioned at the bottom rim of the scaffold, spinning flat non - woven silk patches as they locally reinforced the gaps across CNC - deposited silk fibers.

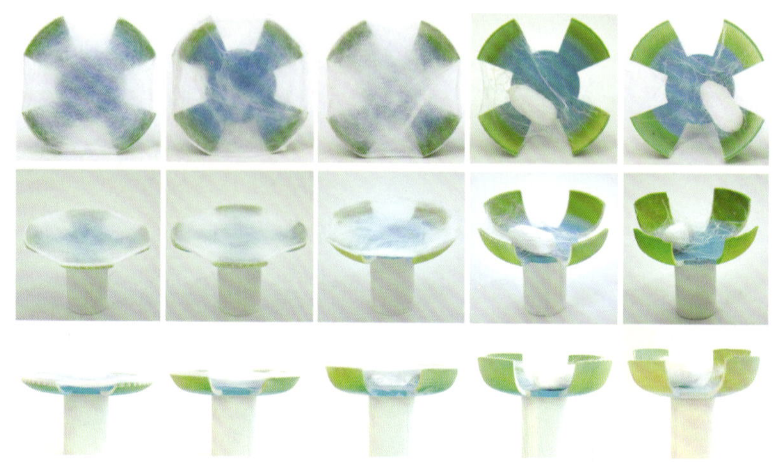

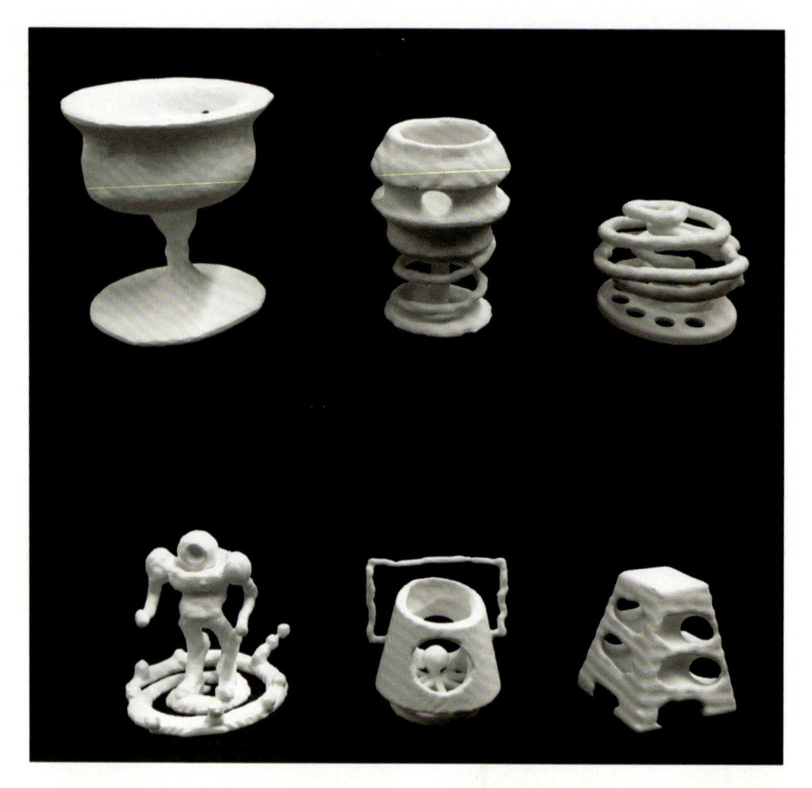

在三维中画草图 / Sketching in 3D

草图被当作一种高效的方式用来探求设计构思。这个方案探索并提出一种新的三维草图计算机辅助方式，这吸取了现实与虚拟世界共同的优势。两种工具原型被发展和比较。第一种原型采用运动侦测传感器，投影屏幕以及动作追踪软件。使用者双手的运动变成直观的界面在虚拟空间中三维塑型。第二个原型采用一种手持设备和基于标记的增强现实技术。这种手持设备从所需的角度显示虚拟物体并作为虚拟的工具来工作，可以在三维空间中建立虚拟物体。

Sketching has been used as an effective way to explore design ideas. This project investigates and proposes new computer-aided methods of 3-dimensional sketching, which take advantages of both the physical and virtual worlds. Two prototype tools were developed and compared. The first prototype uses a motion detecting sensor, projection screen, and gesture tracking software. The movement of the user's hands becomes the intuitive interface to shape 3-dimensional objects in the virtual space. The second prototype uses a hand-held device with marker-based augmented reality technique. The hand-held device displays the virtual object from desired angles and works as a virtual tool to shape virtual objects in 3-dimensional space.

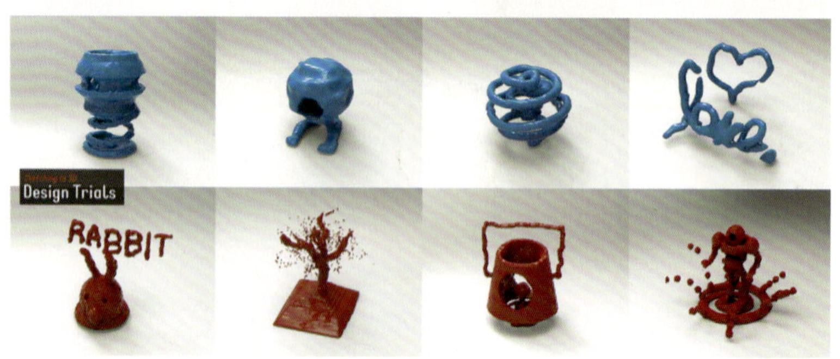

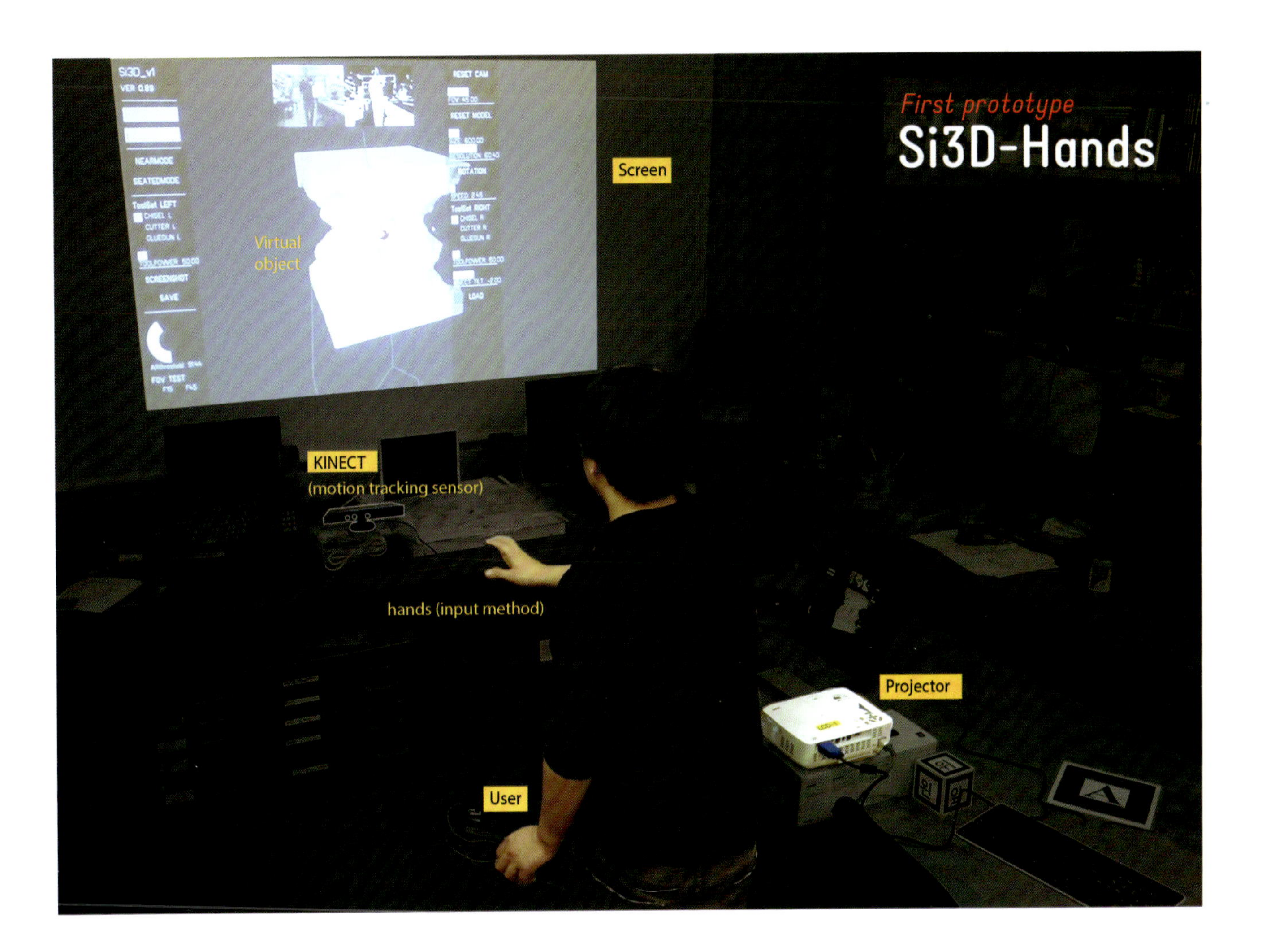
First prototype
Si3D-Hands
Screen
Virtual object
KINECT
(motion tracking sensor)
hands (input method)
Projector
User

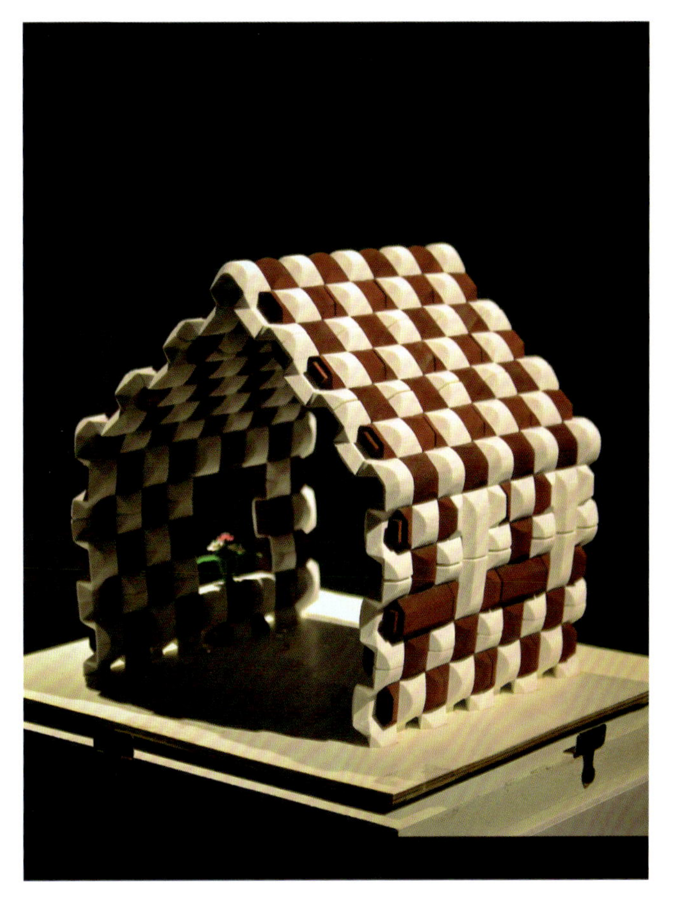

后砂浆建筑 / Post-Mortar Architecture

这个方案运用形态语法的方式，计算性的运用新型制造体系来阐释传统的编织技术。组装“编织砖”同传统的手工艺人织线逻辑相同。通过连接和组合两种不同颜色的砖，这个体系允许手艺人或者设计师从结构元素来表达任意二维图案。受传统珠工艺的启发，“珠砖”模块的抗压连同线的张拉，共同演绎了后张拉结构体系。装饰性的“珠砖”可以被排列进二维组合来表达不同的主题，或者组合为三维的结构性构造来建立多种建筑形式。

These projects use Shape Grammar methods to computationally interpret traditional weaving techniques into a novel fabrication system. WovenBricks are assembled in the same logic as the traditional weaver weaves the threads. By connecting and composing two different colors of brick, this system allows craftsmen or designers to express any binary motifs as a structural element. Inspired by traditional bead craft, the compressive BeadBrick modules, together with the tensile threads, perform a post-tensioning structural system. The ornamental BeadBrick can be arranged into two-dimensional composition to render various motifs, or into 3-dimensional structural configuration to build many architectural forms.

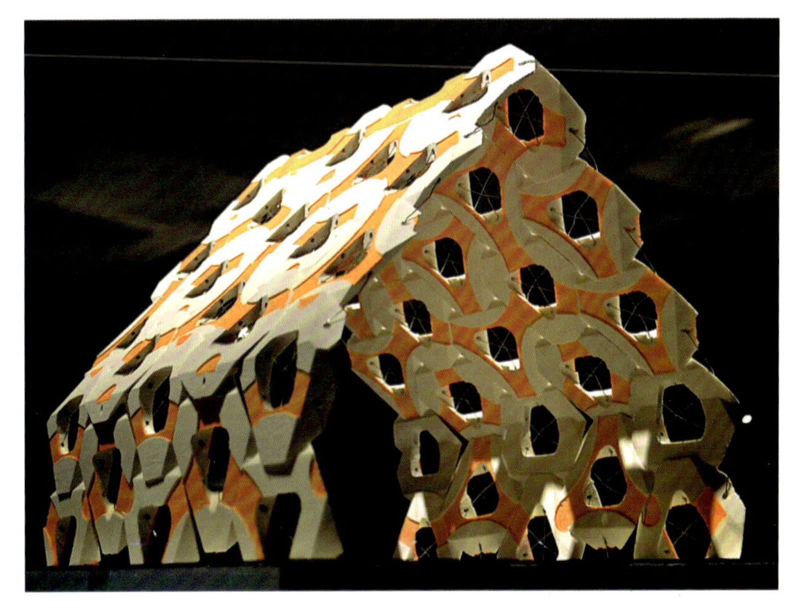

美国普瑞特艺术学院 / Pratt Institute

塑造线条 / Fashioning Lines

这个题目通过一个立足线性变化的设计项目体现了一系列关系，包括建筑与构造、建筑与材料、建筑与两边差别的关系。边界线和矢量线曾经是建筑建设基础，如今被重新设置和设想，对个体与群体产生持续神奇的影响。本次设计题目建议在曼哈顿西部和海兰恩之间建立一系列时尚的公共场景空间。

This project plays out a series of relationships in a studio addressing variation in linear intensities. These relationships include those between architecture and fabric, architecture and textiles, and architecture and the lines of distinction between one side and another. Boundary lines and vectoral lines of flow - which were once the stable production of architecture – are redeployed and reimagined to produce often strange and estranging affects in an ongoing and evolving play of imagined individual and collective bodies. In this studio, a proposal was developed for a fashion collective space for sites between the western edge of Manhattan and the Highline.

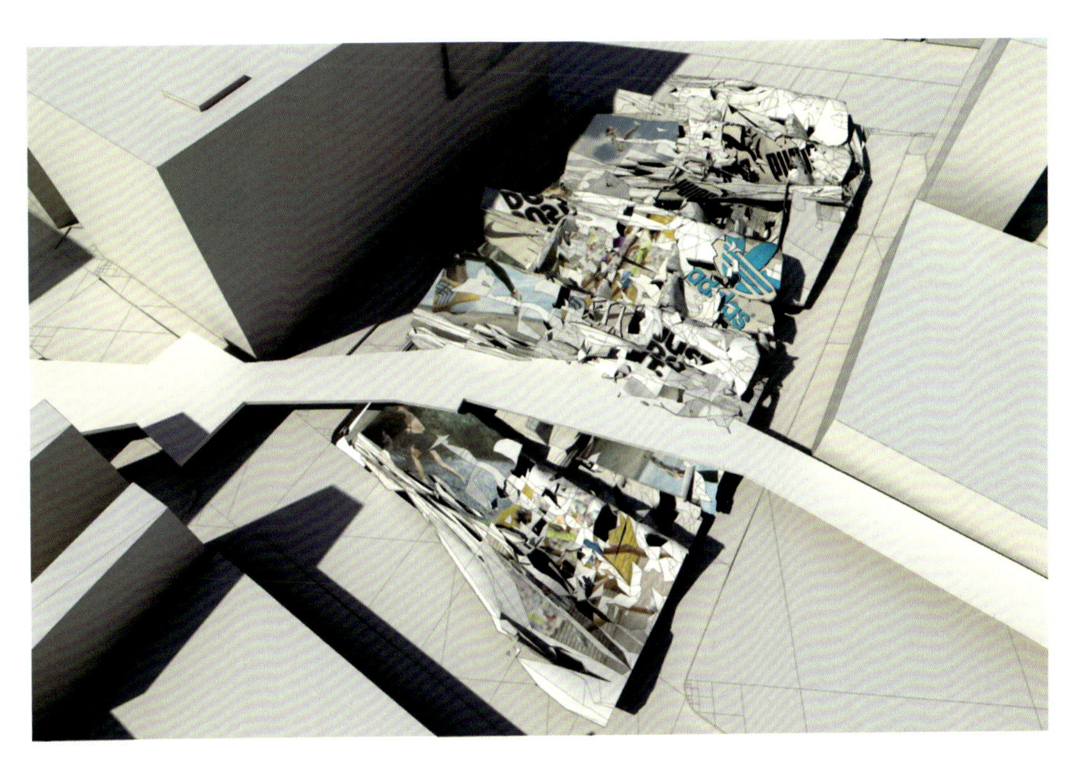
adidas

第六个镇区 / The 6th Borough

这个探究性的城市设计题目研究了纽约城人工地质的持续变迁。除了照旧尝试恢复 21 世纪城市原有的自然秩序之外，纽约城的都市现象继续体现着自然与人工看似奇怪的合作中拥有的惊人潜力。曼哈顿、布鲁克林、布朗科斯、皇后区和史丹顿岛是曼哈顿地区五个著名的镇区。这个题目假想了一个“第六区”，它一直是纽约城周边的水路，需要仔细分析新的人工地质的潜力，探究用于港口设计的城市模式。

This speculative urban design project explores the continuing transformation of the artificial geology that in New York City. Despite nostalgic attempts to recover original natural orders in the 21st century city, urban phenomena in New York City continue to demonstrate the astonishing potentials of strange collaborations between artificial and natural regimes. Manhattan, Brooklyn, the Bronx, Queens, and Staten Island are the famous five boroughs of Manhattan. This project locates as a site a hypothetical "6th Borough" that has always been New York City's surrounding waterways and examines the potential of new artificial geologies and urban patterns that can be designed for the estuary.

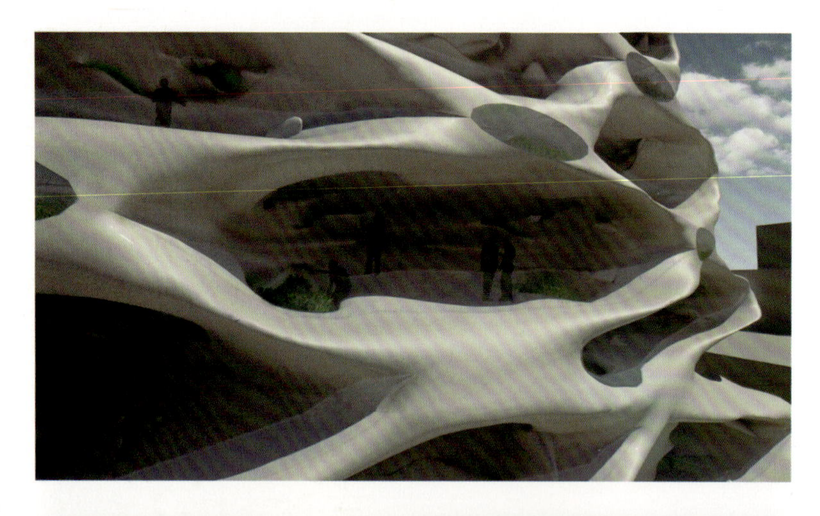

耕作极限 / Farming Extreme

耕作极限这个题目关注针对日益凸显的垂直农业项目的新建筑类型的发展。这个项目引发了对都市风格和农业方面垂直性的再思考。反过来，它也引发了对农业的幻想：和地面没有联系，可以脱离土地生长。无土栽培利用氢氧化物提供营养，这在城市种植活动中占有很大比例。脱离地面获得自由，蔬菜可以变成关注的对象——更大空间中的设计元素。

Farming Extreme is a project that focuses on developing a new architectural building typology in response to the emerging program of the vertical farm. The program suggests a rethinking of urbanity and verticality in connection with agriculture. Inversely, it also constitutes a re-imagining of agriculture that is not connected to the ground and that grows without the presence of earth. This un-soiled way of growing operates through the use of hydroponics and constitutes the majority of urban farming activities. Freed from their connection to the ground, vegetables are allowed to become objects of contemplation –design elements within larger spatial installations.

美国普林斯顿大学建筑学院 / Princeton

跨越布鲁塞尔：行动的理由 / Transnational Brussels

这个方案考虑到建筑师在跨国语境下的政治代言。关注布鲁塞尔这座城市，作为跨国空间的典范，该方案提出了一个策略性的反思和记录，关于那些可为政治活动容纳临时形式的城市空间。从布鲁塞尔 Leopold Quarter 的总平面图可以看出，新的公共空间介入到密集的已建成的邻里单位，通过一系列大量的方案活跃了新划分的城市空间。这些“介入体”坐落在不同权力轨道间的重叠和壁龛中，在那里它们就如同空间污染和政治颠覆的工具般运作。

This project considers architecture's political agency in a transnational world. Focusing on the city of Brussels as a paradigmatic transnational space, the project proposes a strategic rethinking and recoding of urban space to accommodate contemporary forms of political action. The masterplan for Brussels' Leopold Quarter includes the insertion of new public space into this densely built-up neighborhood, alongside a wide spectrum of program that animates the newly claimed urban space. Interventions are situated in the overlaps and niches between different orbits of power where they operate as instruments of spatial contamination and political subversion.

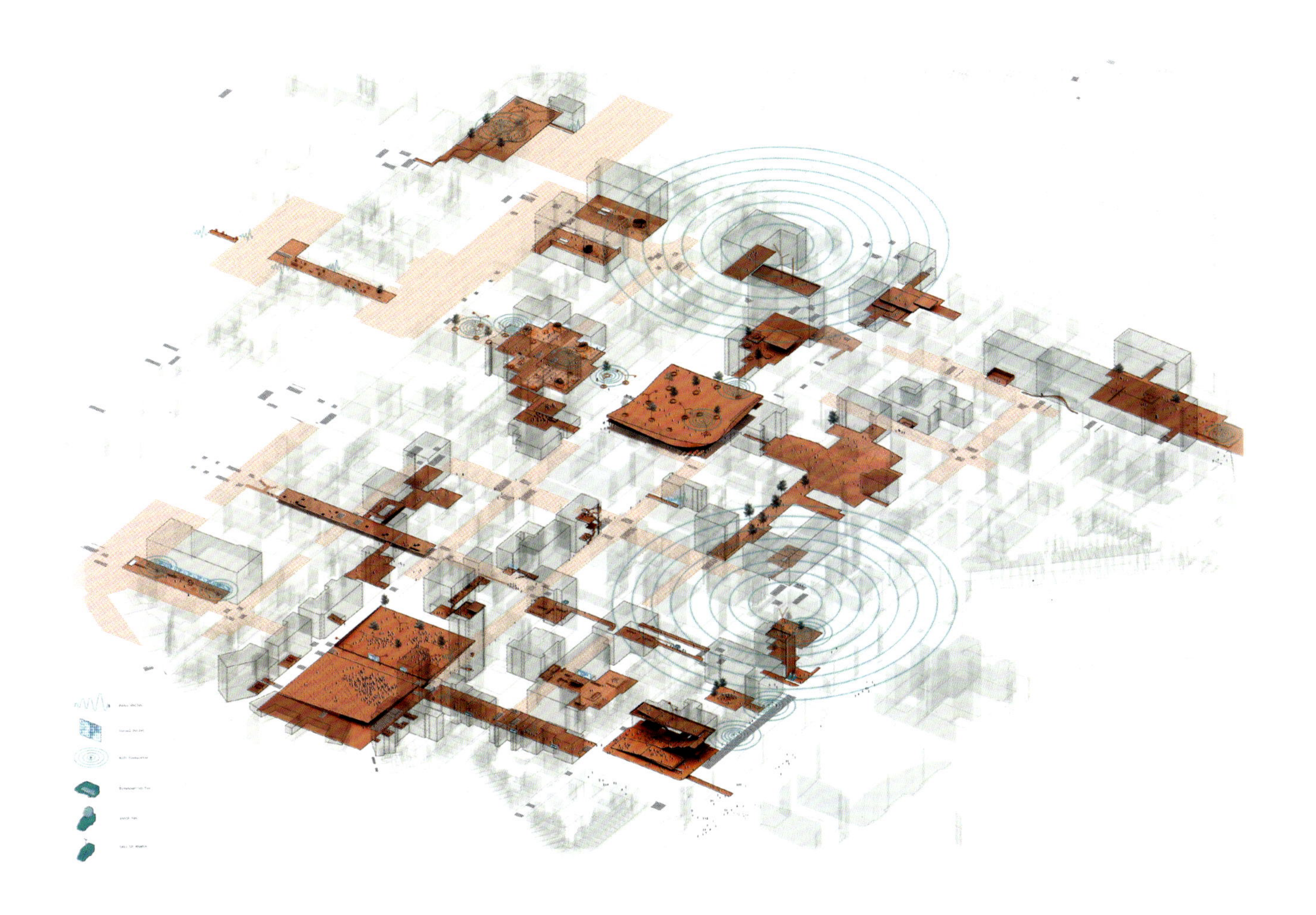

从声音响起 / From the Sound Up

我们如何从对声音的偏好来生成形式？建筑师可否为音乐设计空间，以其听觉本质而不是猜测它们看起来像什么？通过软件“SoundUp”以及与音乐家在空间声学上的合作，建筑师们可以开始其设计过程。一个直观的界面允许使用者去“绘出”他们对声音的偏好，并实时倾听声音模拟的结果。而后建筑师可以通过设定参数限制条件（范围从场地条件到形式语言到建造限制）来指导生成相应的几何围护体。

How might we generating form from acoustic preferences? What if architects could design spaces for music starting with how they sound rather than guessing how they should look? With SoundUp software, architects can start the design process by collaborating with musicians on the acoustics of a space. An intuitive interface allows the users to “paint” their acoustic preferences, hearing the results in real-time audio simulation. The architect can then guide the generation of a corresponding geometric enclosure by setting parametric constraints ranging from site envelope to formal language to constructional limitations.

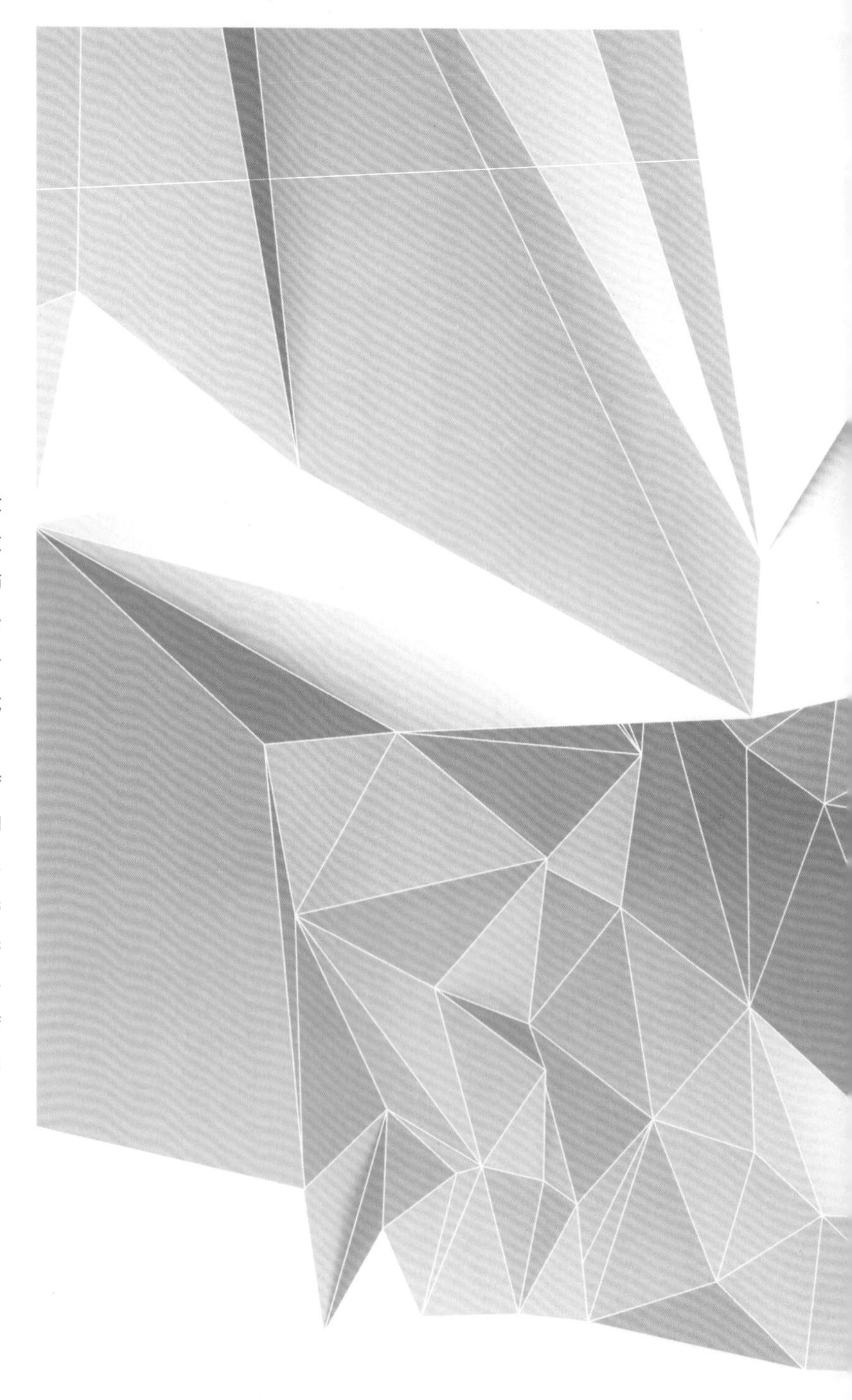

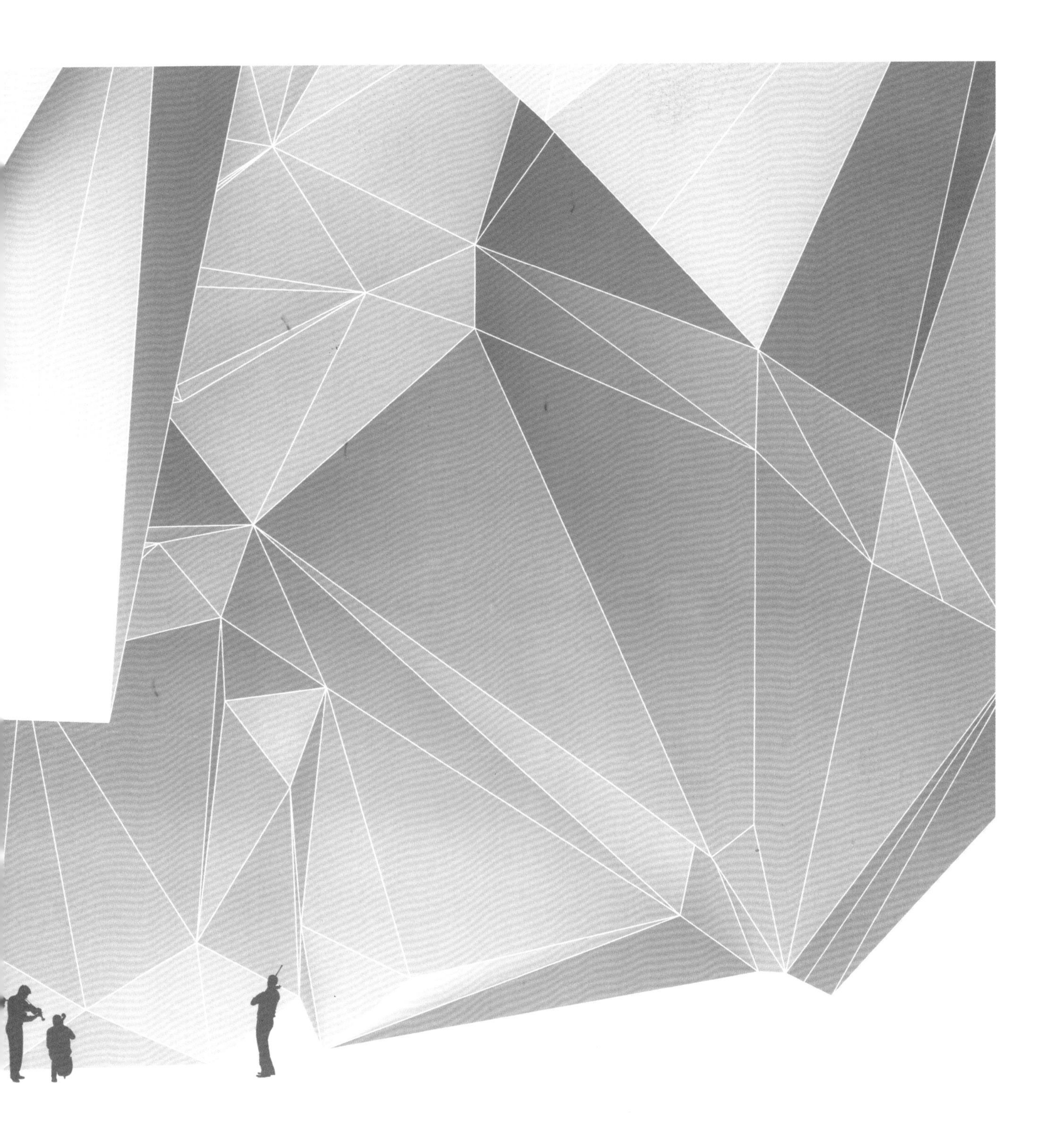

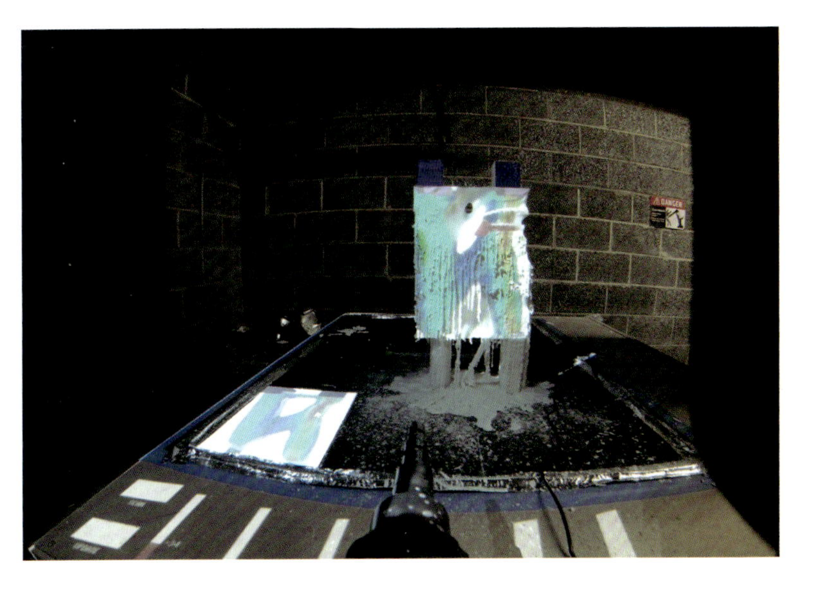

杂糅现实的建模 / Mixed Reality Modelling

这个方案是一个设计—制造的训练，过程结合了物理上的人工输入和机械上带有随机性的材料操作（熔蜡）。溶解一个实体并快速扫描其过程，基于给定的蜡质体量和用户设置的读取条件，系统会尝试获取一个拓扑上优化的结果。众多不断互动的变量给人们操作带来实时性，如对计算机模拟、机械臂的操作以及材料变化过程得控制。这些元素变得不可分割，从全局系统看来它们单独的输入变得难以区分。

This project evolved as a recursive design-fabrication exercise which combines physical human input with the robotic manipulation of a stochastic material process (melting wax). By rapidly scanning a physical object while also melting it, the system attempts to achieve a topologically optimized result based on the given wax volume and user-placed loading conditions. A multitude of constantly-communicated variables give simultaneous control to the human operator, the computer simulation, the robotic manipulator, and the material process. These elements become inseparable, and their individual import becomes indistinguishable from that of the global system.

美国伦斯勒理工大学 / RPI

丛 / Plexus

在本次设计中，学生探索了一种程序的细分方式，作为基于一个简单的笛卡尔网格生成复杂空间形体的方法。每个学生发展出一种算法，实现与网格的互动，产生出各种所谓的"空间行为"，包括有周期规律的以及没有周期规律的。设计的本意是重新思考空间组织，将其视为一种离散的操作，在进行这种操作时，通过不断聚集产生出一种新的组合结果。学生用不同的计算机设计工具以及不同的材料通过数控技术实现他们的设计。

For this design studio students explored procedural subdivision as a means of generating a complex geometric space from a simple Cartesian grid. Each student developed an algorithm for interacting with the grid with the intent of producing a variety of what was called "spatial behaviors" ranging from periodic to aperiodic. The aim of the studio was to rethink spatial organization as an accumulation of discrete operations that aggregate to produce a new range of combinatorial results. The students used varieties of computational design methodologies as well as material experimentation using computer controlled fabrication technologies to realize their designs.

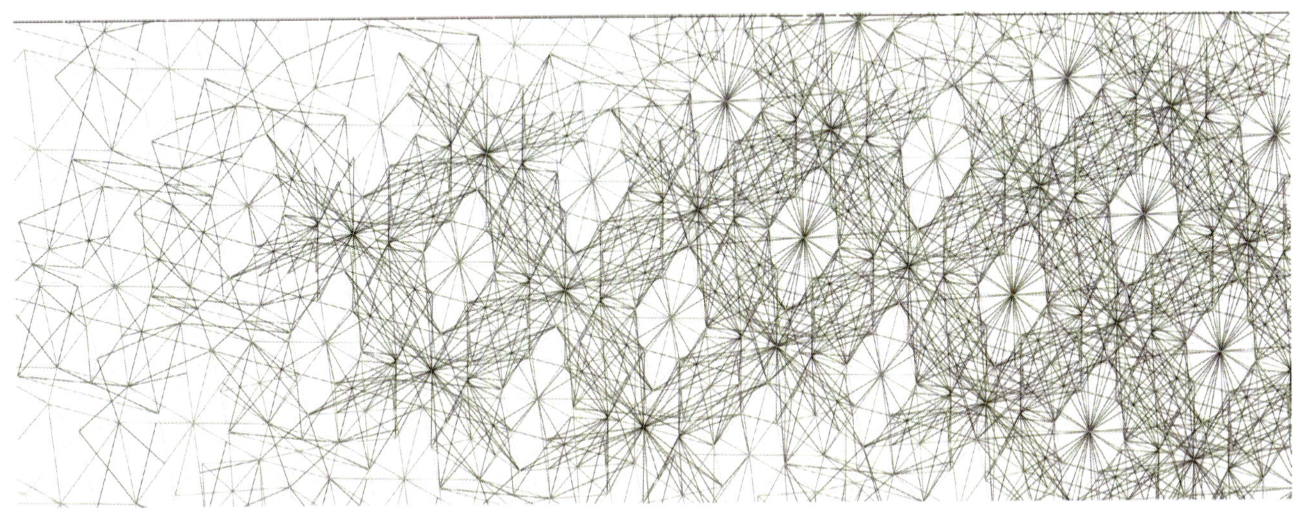

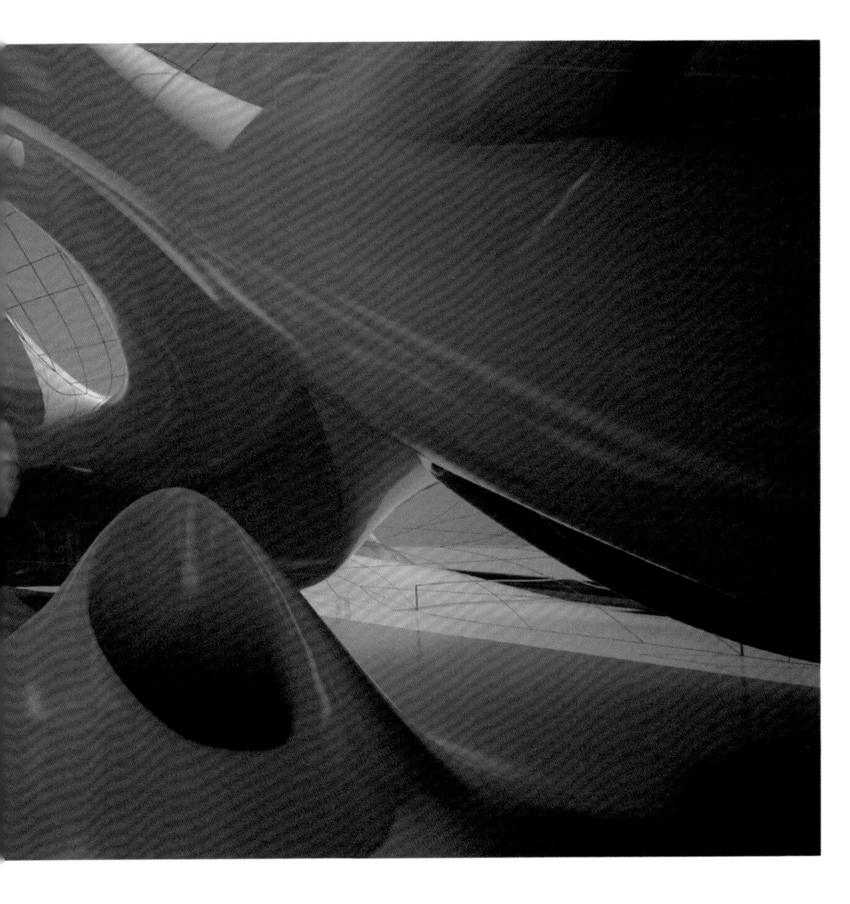

规范的流动性 / Normative Fluidity

本次设计的长远目标是试图在多个领域，如心理、政治、人类学、设计以及建筑，来描述物体怎样通过更直接的场景互动来影响社会。本次设计选择了一个科学博物馆，重点关注科学现象的感官体验作为设计的概念来源。它从操控灯光开始。每个实验通过包容、塑形以及操控都会产生独特的、确定的灯光效果。数字技术提供了一个新的机会，形成从标准化到个人定制、再现到模拟、静态到动态、减少到增加、固定死板到柔软可变的转变。

The term affect is increasingly being using in many fields including psychology, politics, anthropology, design and architecture to describe how objects can influence society through a more immediate engagement of the senses. This studio for a museum of science focused on transposing sensorial affects of scientific phenomena as a generative concept for design. It began by drawing and modeling with light. Each experiment produced distinct and tangible light affects through containing, shaping, and manipulating light. The digitally repositioned affects afford new opportunities through advanced design techniques transitioning from standardization to customization, representation to simulation, static to dynamic, subtractive to additive, and rigid to pliable.

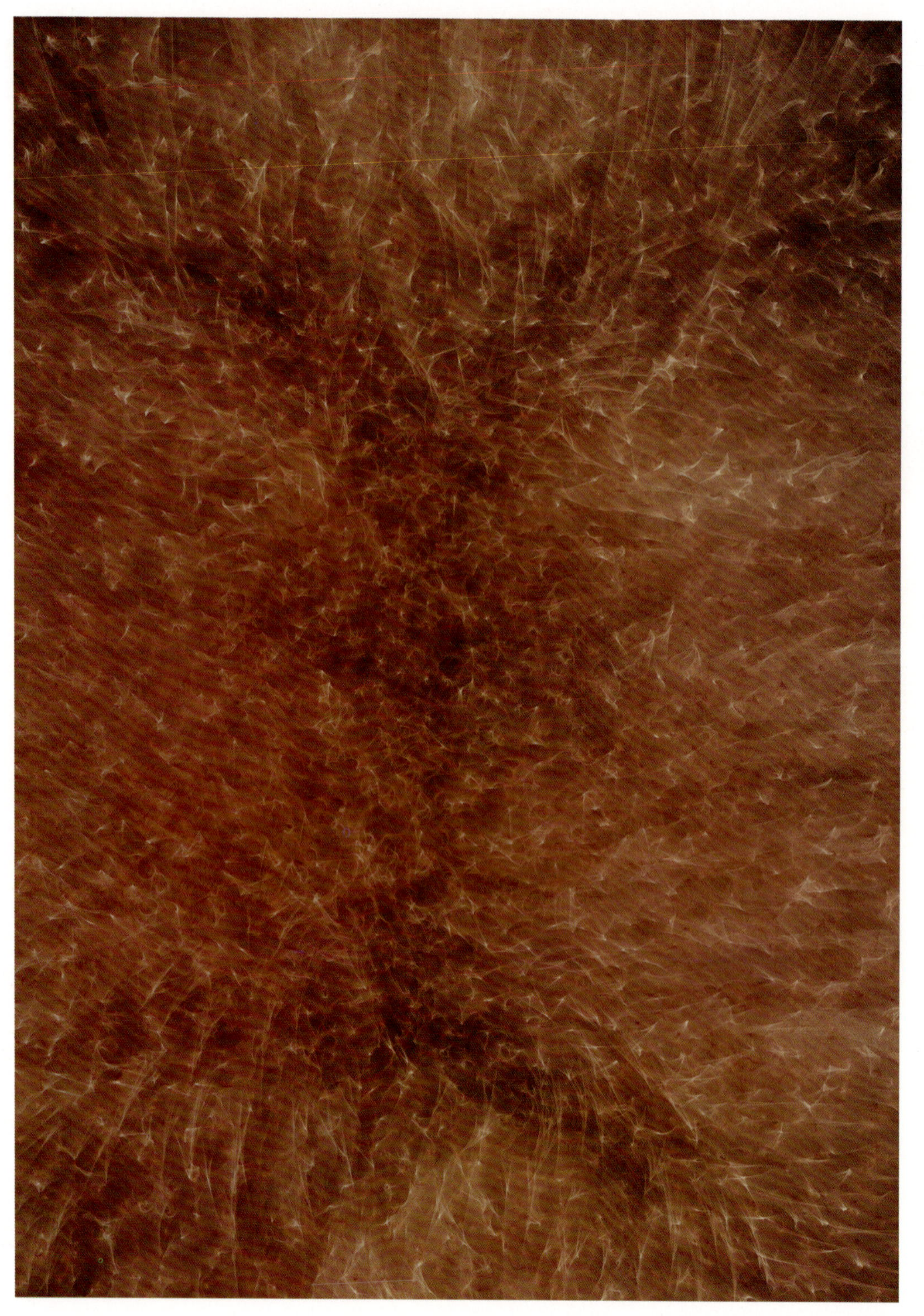

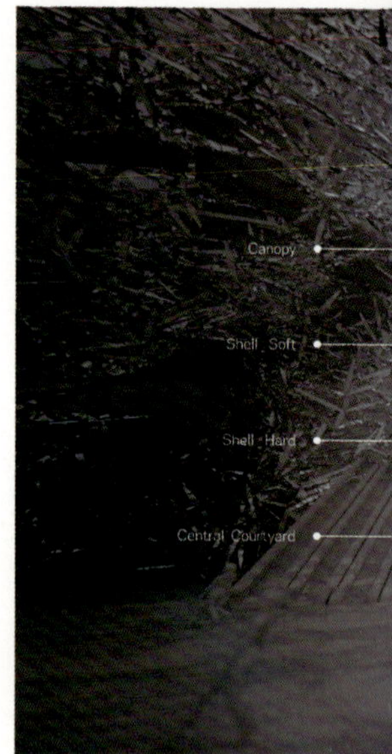
Canopy
Shell Soft
Shell Hard
Central Courtyard

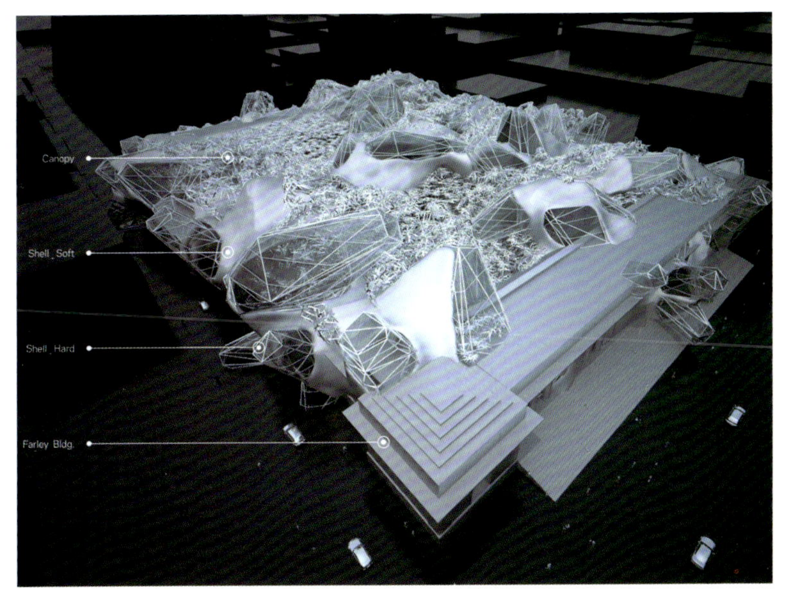

情感机器 / Affective Machines

本次设计重点研究复杂适应系统的潜力，看它同时组织和定义人类活动空间以及异质的实验条件的能力。学生们在纽约的 Penn 车站实现这些系统。这些从多代理系统中发展出来的行为是从 19 世纪荷兰的静物以及维多利亚时期的自动机器人中提取出来的，它们被魔术师用来创造一种具有美学效果的系统，这种系统与自组织的仿生系统相对立。这一系统可在不同的建筑尺度以及不同的层级组织中运行。

This studio focused on researching the potential of complex adaptive systems to simultaneously organize and define the space of human occupation as well as address issues of heterogeneous experiential qualities. Students implemented these systems in the design of a new Penn Station in New York. The behaviors developed for multi-agent systems were extracted from the analysis of 19th century Dutch still-lifes and Victorian automatons used by magicians to produce systems aimed towards the production of aesthetic effects as opposed to a bio-mimetic system which preferences self-organization. The systems are designed to operate across multiple architectural scales and hierarchies.

美国南加州建筑学院 / SCI-Arc

用石头雕刻 / Carved in Stone

本项目旨在扩展数字表皮的潜力，在建筑表皮的设计和建造中开发神奇的光学和触觉感知。用了实物三维扫描，虚拟模型 ,Zbrush 数字编辑，计算机数控技术等一系列复杂的设备和步骤，这个项目希望以此扩展计算机技术的应用范围。由于它可以数字操纵最初对石头的三维扫描，在木头上添加经修改的石材纹理，向自然材质中注入合成材料，因此，石头雕刻模糊了真实的和人工之间的界限。

This project aims to broaden the potential of digital surfaces and provoke novel optic and tactile sensations in the design and fabrication of an architectural envelope. Using a complex array of equipment and procedures, from three dimensional scanning of natural objects, virtual modeling, digitally edited paint in Zbrush, and CNC machining, this project attempts to stretch the boundaries in computational techniques. Additionally, Carved in Stone sets out to blur the distinction between what is authentic and artificially constructed through digitally manipulating the initial 3D scans of stone, superimposing the modified stone textures onto figural wood, and injecting synthetic material into natural matter.

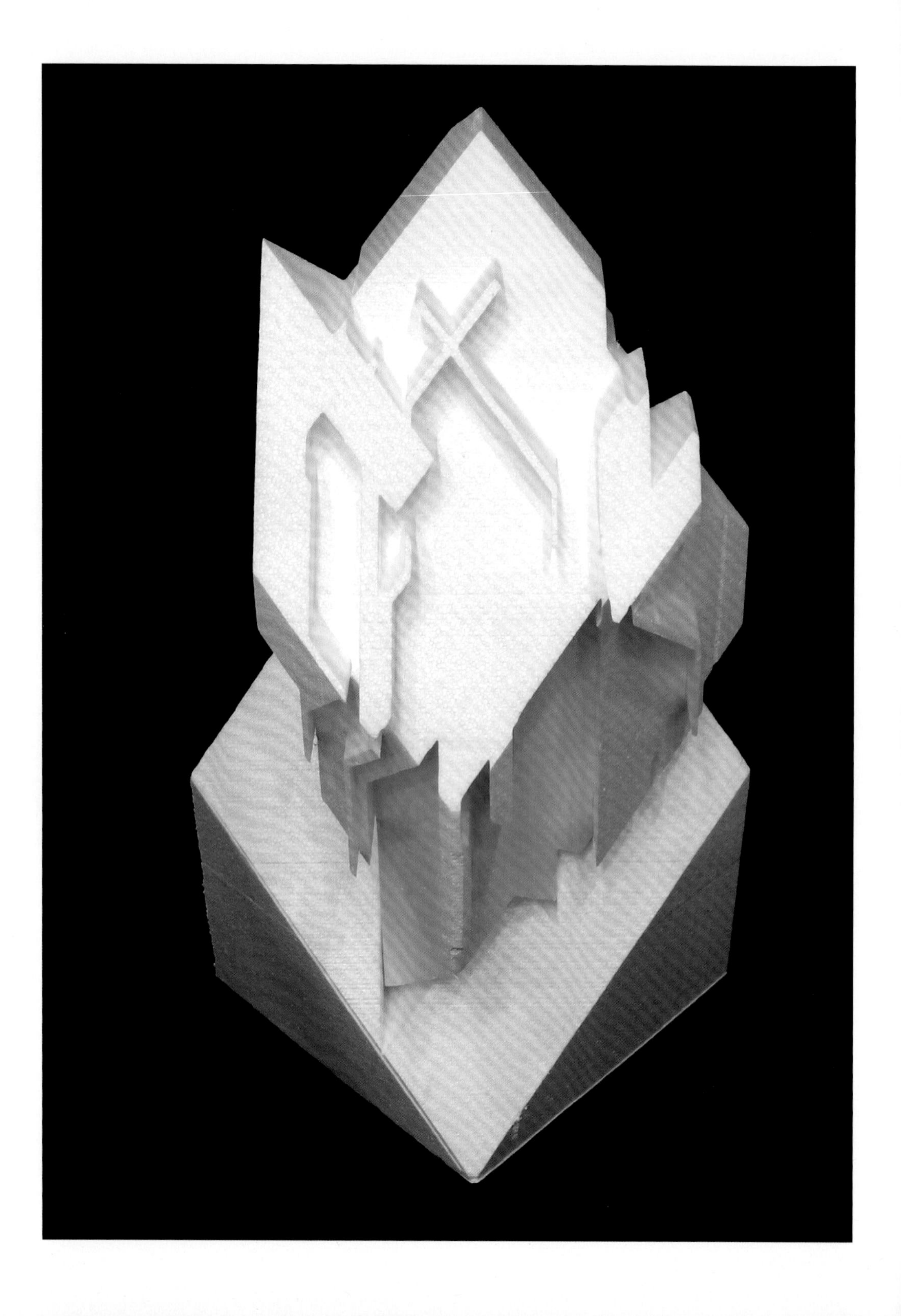

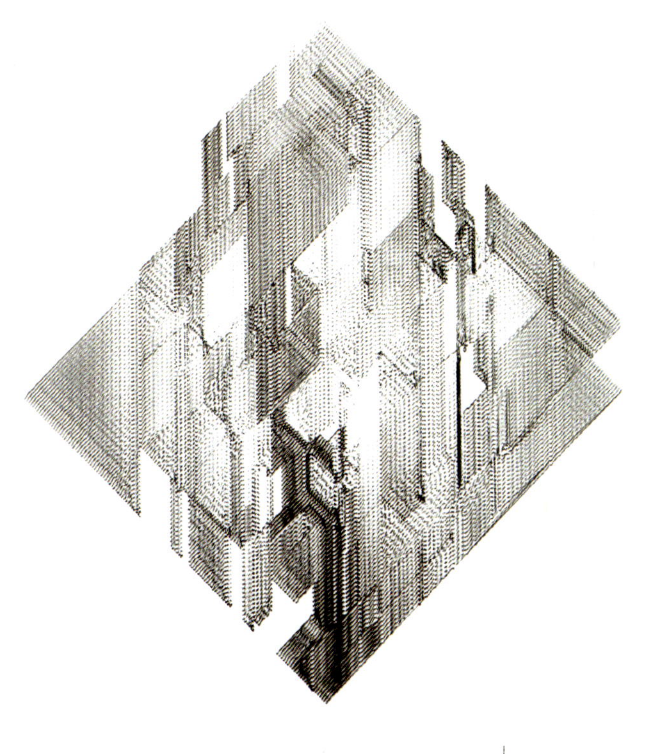

低保真度 / Low Fidelity

低保真度以一个建筑对象为起点，经过转译、抽象，最后呈现出一幅图画。一幅图画是一个二维的展示，具有变形、抽象的特点，和原始对象具有一定差别。这个项目关注的就是图画与物体之间的差别，对我们曾经习惯的具体化提出挑战。题目要求记录空间中的特定点，简化曲线，通过六轴移动控制和机械弱挤压来将得到的形式具体化。物体是什么样不重要，这样就开启了一个讨论和设计的临界空间。

Low Fidelity begins with the architectural object. It moves through translation and abstraction to its representation: the drawing. While the drawing continues to be a two-dimensional representation with the ability to flatten, abstract, and distance itself from the object, and to produce friction between objects and their representations, the project engages the infidelities that challenge our interface with materialization. Through the registration of projected points in space, the reduction and delineation of a single curve and the materialization of the drawn form through 6-axis motion control and mechanically attenuated extrusion, it alienates objects of representation to open up critical space for discourse and design.

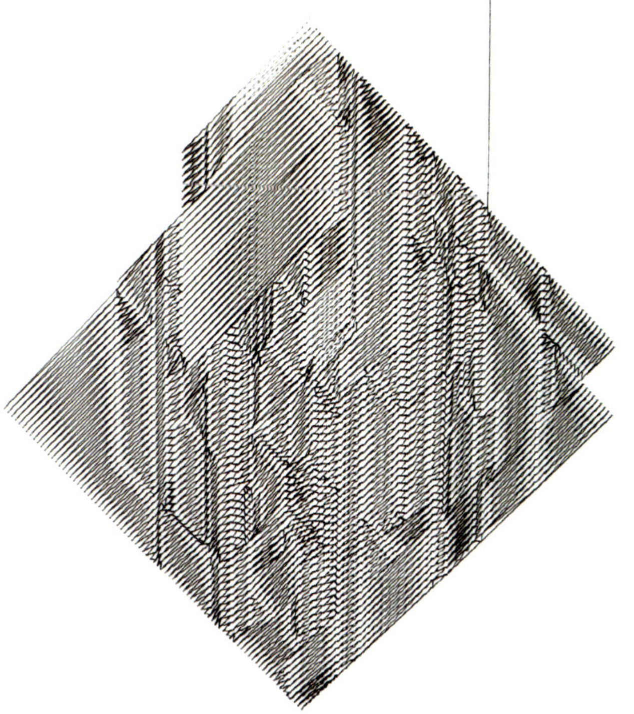

“脆弱”的线条 / Vulnerable Lines

这个项目旨在实现用如今更有效率的工具来进行建筑作图。这些工具带来的阅读体验对建筑有重要的意义，并产生了一种新的建筑实体和事件。计算机设计的线条可以分层拼贴。新加入的编码色彩的图层赋予这些线条艺术的意向，在各种媒介中进行数字化虚拟线条时引入了维度。图像化通过这些过程重新融入了绘图当中。相互交错的线条、色彩、材料等图层和虚拟技术都会导致误差的产生。这就在满足广泛阅读的同时为绘图增添了神秘感。

This project aims to achieve architectural drawings with the generative tools of today, where the reading performance they elicit become architectural significations in their own right, producing a new kind of architectural entity or event. Computationally designed lines are layered and collaged. While the added layers of scripted colors contaminate these lines with artistic intentionality, dimensions are introduced by digitally fabricating these lines on various media. Reinserted into the drawings through these processes is the pictoriality. Multiple levels of misregistrations take place between the processes, between the interwoven layers of lines, colors, materials and fabrication techniques, adding a sense of mystery to the drawings while allowing manifold readings.

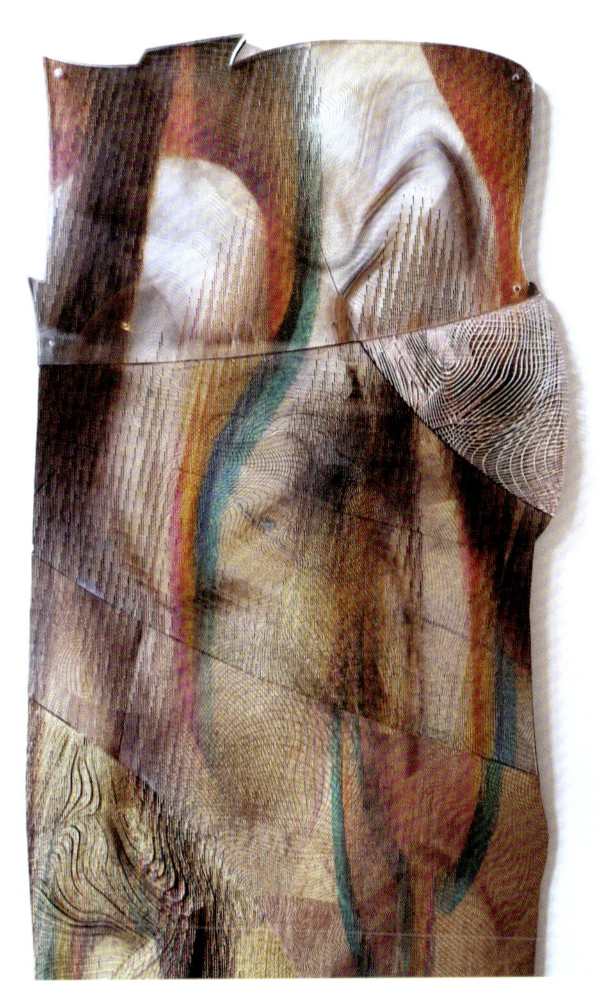

美国加州大学洛杉矶分校建筑系 / UCLA

Mathilde 项目 / Mathilde

本项目试图探索是否可能为一个已经完全自相矛盾的物体增加自身的疏远感。在本项目中，艾森曼的 11A 住宅的形态策略，以及莫里诺的山谷公寓和 Casa Miller 项目中的情感策略被作为重要的设计手段加以运用。该项目利用了 House 11A 中的虚体和实体的组织策略，以及莫里诺公寓中的情感表现手法。褶皱、镜像和玻璃都被作为一种产生视觉错觉的手法，降低了形态的可读性，因而造成了疏远感。这些反射的效果产生了千变万化的幻象，不断波动的抽象视觉，扰乱了观者对形态进行更细致的解读。

The approach to this project was to question whether it is possible to maintain let alone increase the estrangement of an object that is already so completely ambivalent. In this case the formal strategies of Eisenman's House 11A with the affective strategies of Mollino's De Valle apartment and Casa Miller were deployed as first principals. The project institutes the House 11A organizational strategy of solids and voids and the sensual surface strategies of the Mollino apartments. Drapery, mirror, and glass are all objects contributing to the optical trickery that decrease formal legibility resulting in estrangement. These reflective affects result in a phantasmagoria, constantly fluctuating abstractions that frustrate close reading.

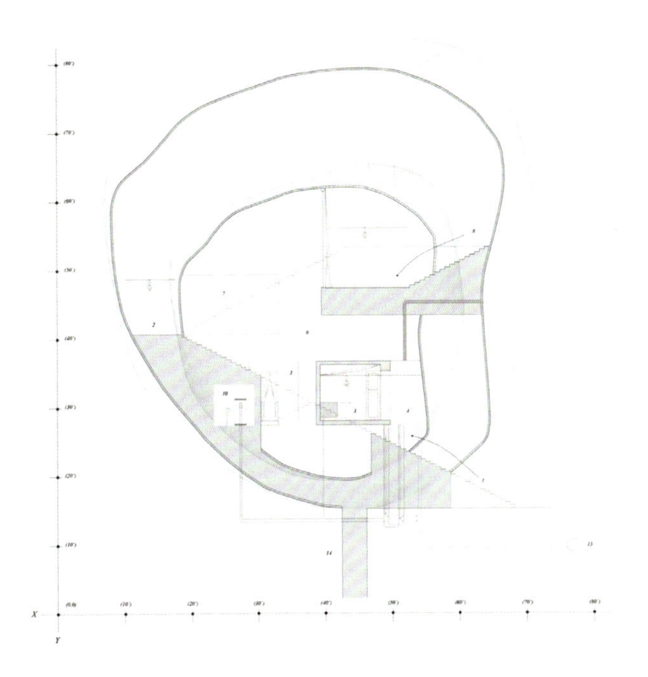

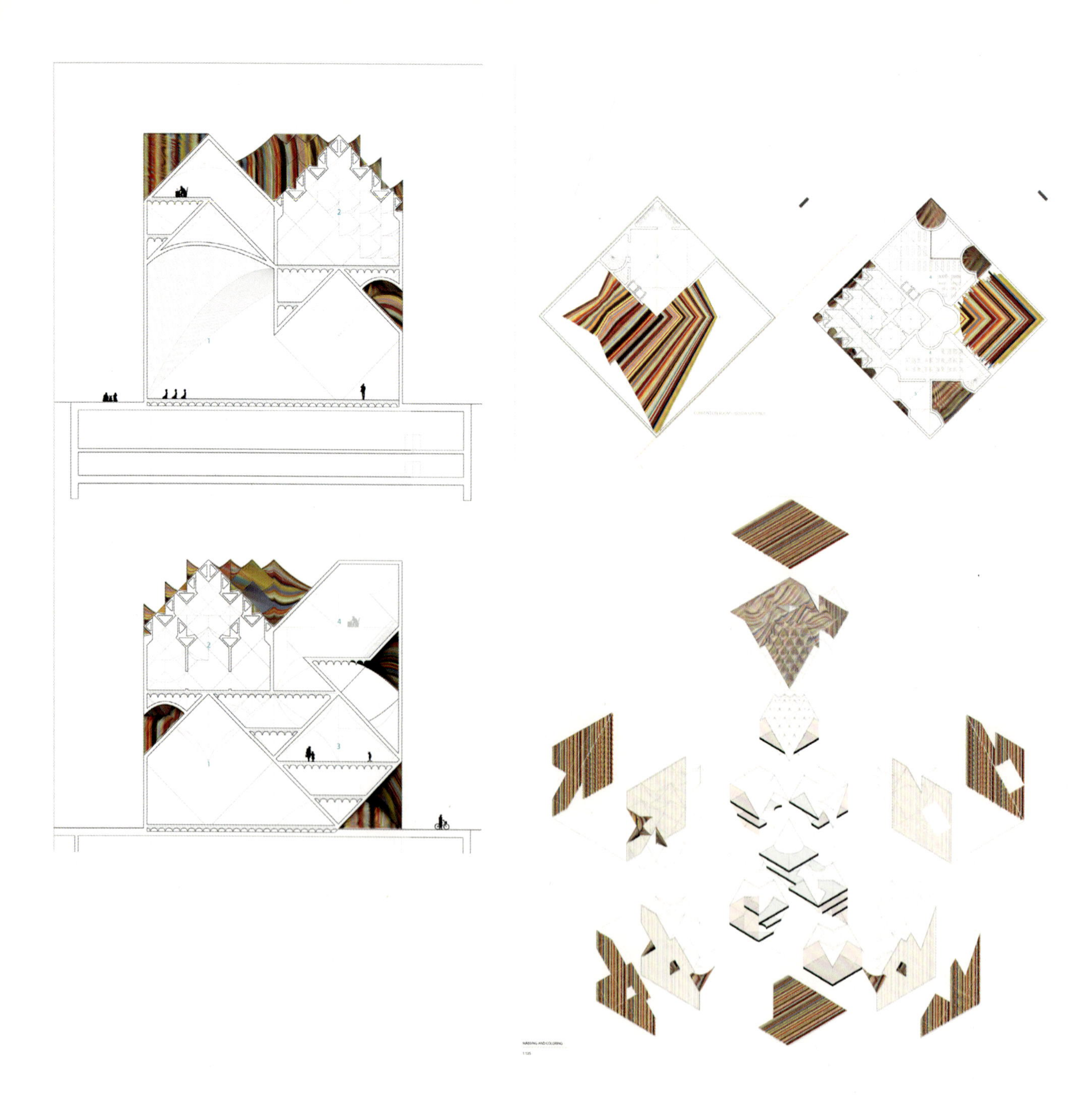
MASSING AND COLORING
1:125

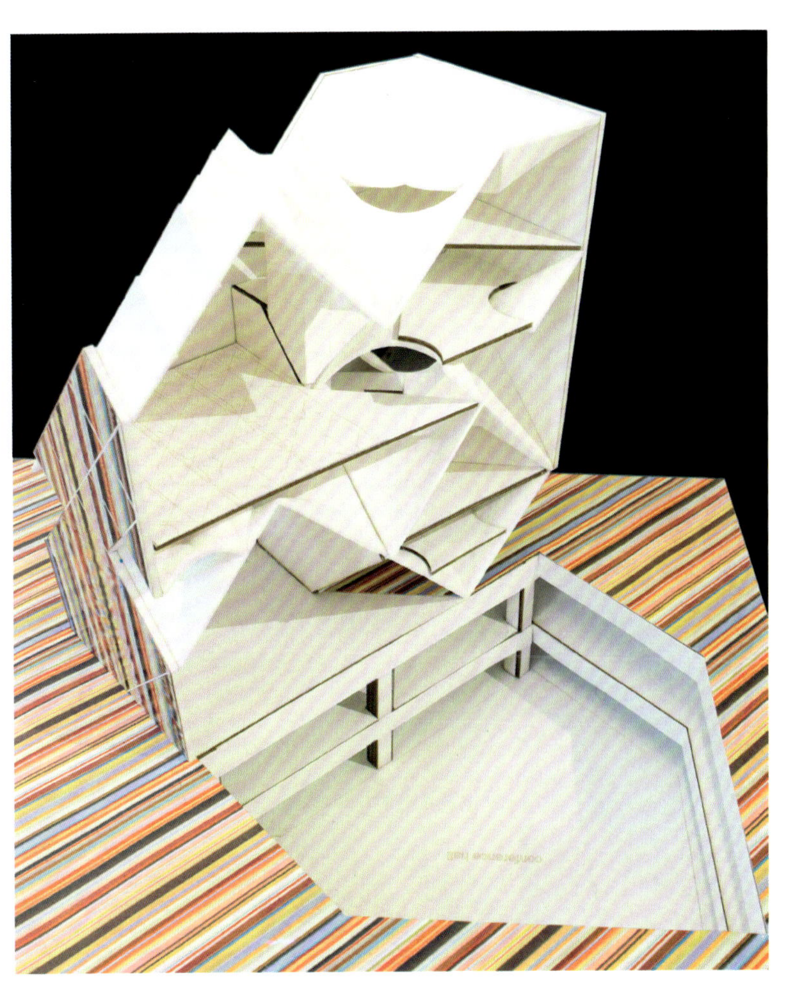

发散的塑制品和静音符号 / Divergent Monoliths and The Mute Icon

历史上，宗教建筑体现出了大胆以及明确的建筑符号，表达政治与社会力量。为了挑战这种策略以及符号的象征性，这个项目对同一建筑中两种相对的范例的关系进行探讨：单一的固体与异质的部分累积。因为符号化是可见的，体块和它的壁纸也要考虑到互补性。这些条纹布的应用——一种绘画的策略，将体块抽象成离散的边，将各部分分离开，同时将整体进行混合，从而进一步重新定位符号的作用。

Historically, religious buildings offer bold and easily readable architectural symbols for the manifestation of political and social power. In order to challenge the formal strategy and representation of such icons, this project explored the delicate relationship between two often opposed paradigms for mass configuration coexisting in the same architectural body: a monolithic solid and a heterogeneous accumulation of parts. Because iconicity is perceived visually, the mass and its wallpaper should also play complementary roles. The application of stripes—a painting strategy which abstracts the mass at its discrete edges—diffuses parts and confuses the whole simultaneously, thereby further redefining the role of the icon.

弯曲的拱顶 / Contorted Vault

12 世纪“lady 小教堂”哥特式拱顶是经过广泛的形式研究分析的。一个新的物件将基于其结构与装饰产生。通过对原始拱顶的分析，分离出了三种类型，从而导致了边、结构构件和装饰间的多样化的理性条件。之后研究了每个小空间中的拉、压、弯以及其他操作。通过这些，产生一个简单化的集成空间，通过 3D 打印、蛋箱模型、切割等高线来进一步表达每一时刻扭曲的物件表皮上清晰与模糊的变化。

The 12th-century Gothic vault at Ely Cathedral's 'Lady Chapel' was manipulated based on extensive formal analytical studies. A new object was produced based on the dialectic between structure and ornament. Analysis of the original vault bay, from which three types were derived, led to a variety of relational conditions between edges, structural members, and ornamentation. The implications of bending, twisting, folding, and other formal operations performed on each bay were then explored. From this a simplified conglomerate bay arose, and 3D prints, egg-crate models, and finely contoured laminates were explored to further inform moments of definition and blurring on the contorted object's surfaces.

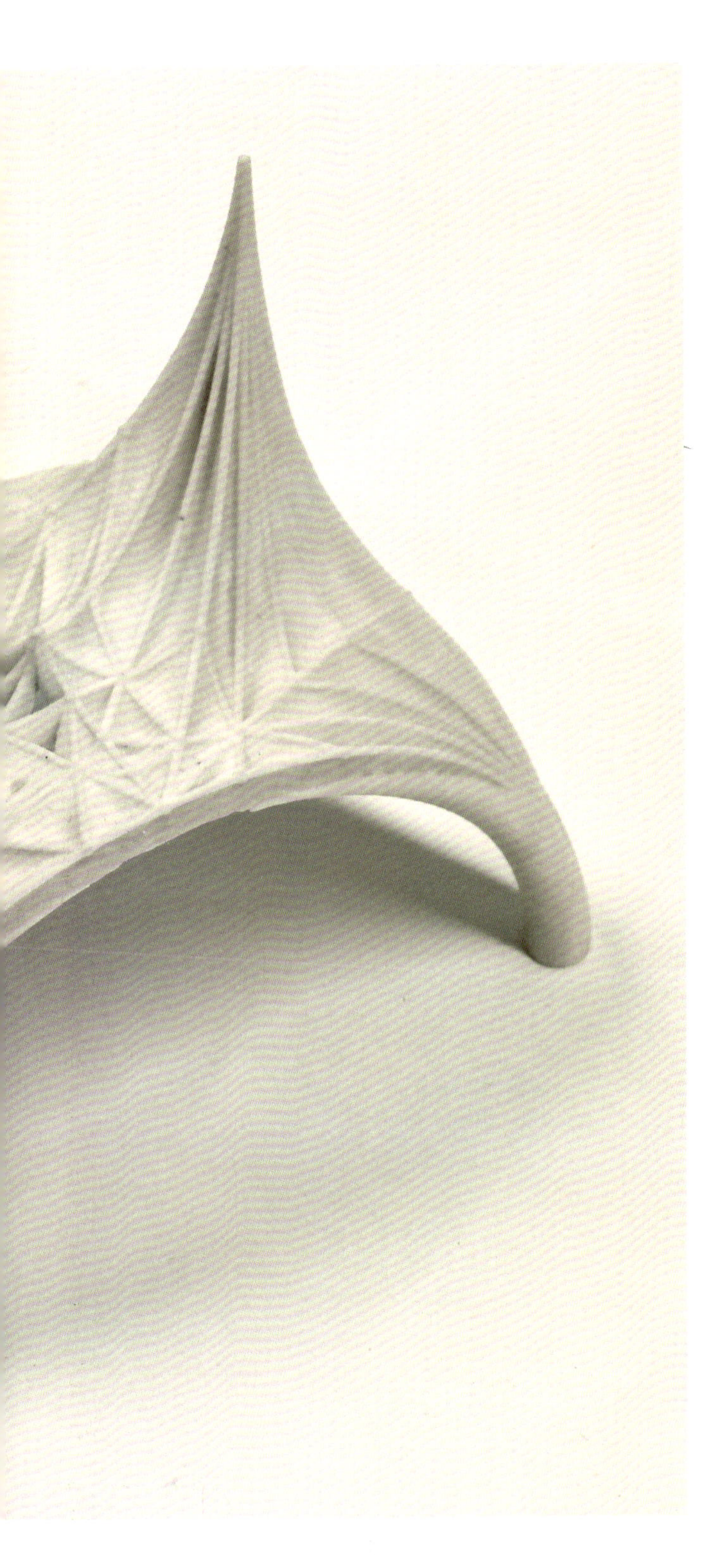

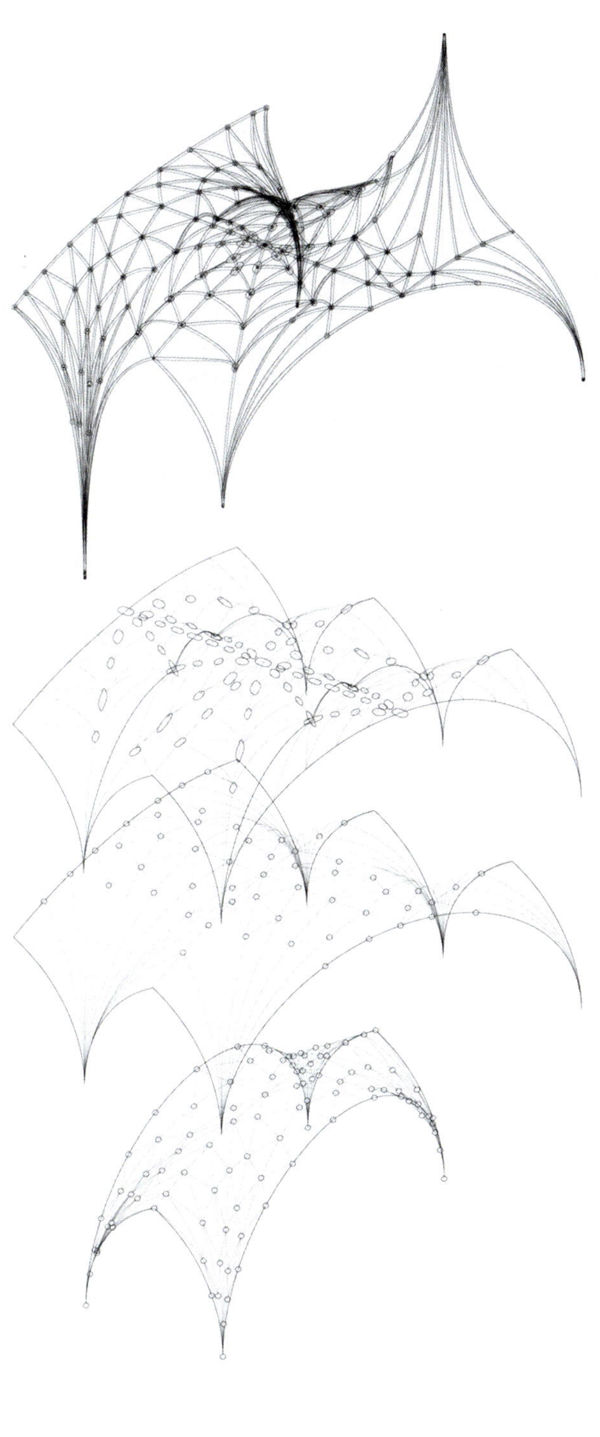

美国密歇根大学 / U Michigan

递归式垂直平台 / Vertical Territories of Recursion

递归式垂直平台是在 Matias del Camp 和 Adam Fure 的指导下进行的设计——制造探索。机器人制造和非欧几里德空间结构的互补揭示了建筑师从直接的空间设计者到间接的自主建设流程管理者的转换。空间不再是通过标准的有逻辑性的平面、剖面和立面来决定，而是转变为依赖工具途径、建设速度和材料流转过程。因此，空间通过一系列的抽象数据集合或者流程分析来构成。

Vertical Territories of Recursion is a design-fabrication exploration took place under the instruction of Matias del Campo, and Adam Fure. The complementary nature of robotic fabrication and non-Euclidean spatial constructs reveals a shift in the role of the architect from a precise designer of space to the indirect administrator of autonomous construction protocols. Space is no longer determined through the standard logic of plan, section and elevation. Instead the construction process is shifted to one that relies on tool-paths, speeds, and flows of material. Thus, space is formed through the playing-out of a set of abstract data sets, or protocols.

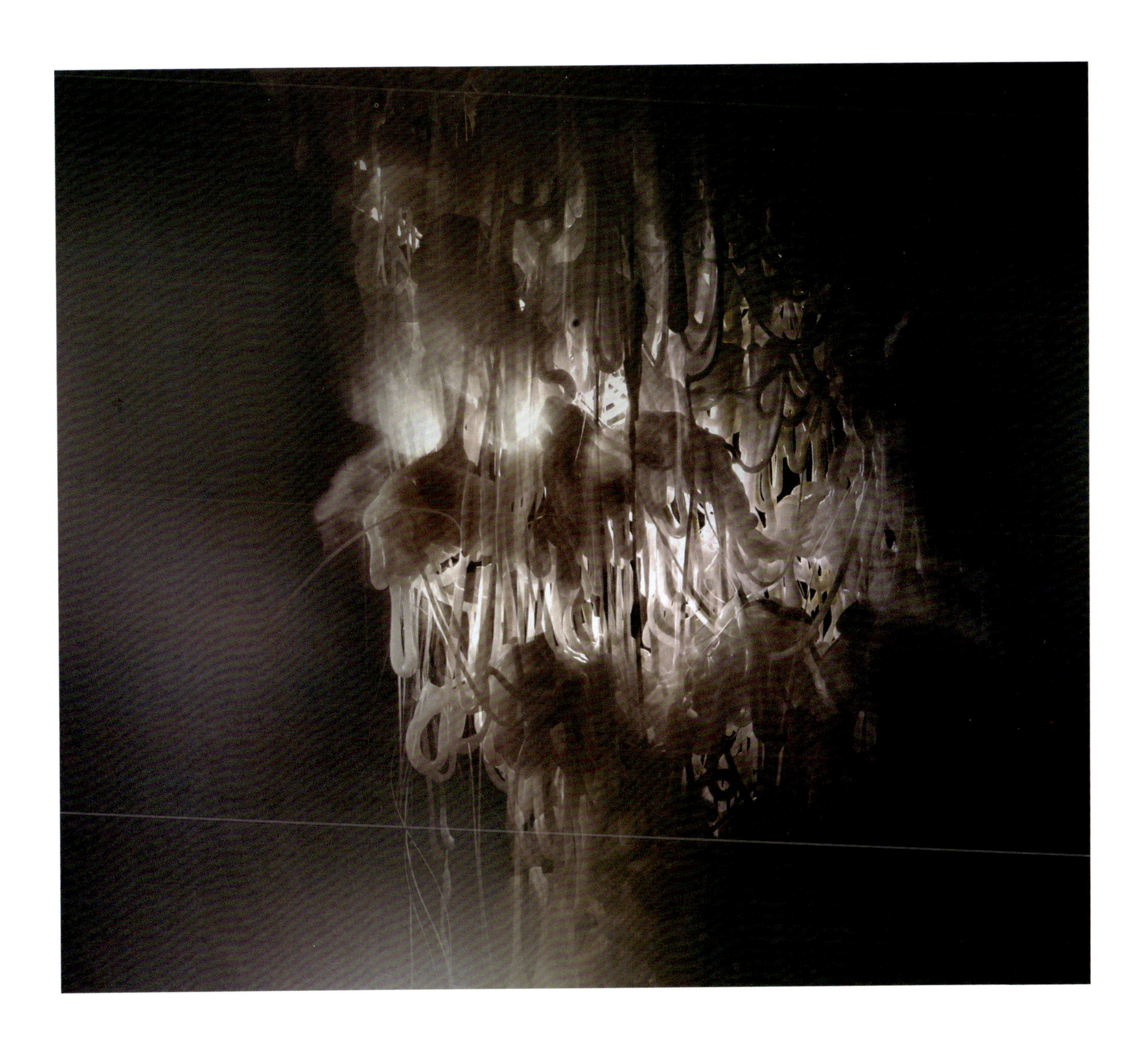

轻质复合材料 / Carbon Wound: Lightweight Composites

该项目利用纤维复合材料和“机器人灯丝绕组”的结合作为一种手段去探索亮度。数字分析和数字脚本可以设计结构，以适应当地的特色材料的属性，其中既考虑到密集型荷载又考虑到发散性荷载设计。“灯丝”的使用作为一种基础材料消除了对标准化规格材料的依赖，并且创造出无材料浪费的自主定制结果。自动化过程引入了高水平的精度和速度来大规模生产可定制的形式以作为一种替代传统制造业和建筑业的方法。

The project utilizes a combination of fiber composites and robotic filament winding as a means to explore lightness. Computational analysis and scripting allow structures to be designed with locally tailored material properties capable of taking into account both the intensive and extensive forces of design. The use of filament as a base material eliminates the dependency on standardized dimensional stock and creates the opportunity for customization with no material waste. The automation process introduces a high level of precision and speed to the mass production of customizable form as an alternative to traditional methods of manufacturing and construction.

活力建筑装配 / Force-Active Architectural Assemblies

该研究探讨超轻量的活力建筑。其特点是通过捆绑配置来稳定平衡结构力以达到巧妙的平衡。进化的材料系统结合了弯曲和张力的组件。在 Sean Ahlquist 处理环境的研究中，材料测试和物理找形研究提供了预测基于弹簧的数字模拟框架。通过一系列使用玻璃钢杆材和聚碳酸酯连接器的实验原型来探索闭环捆绑配置的设计方法。数字化叠加已经在使用 GH 基于原型的局部弯曲半径来预测截面面积方面得到了发展。

This research explores an ultra-light weight Force-Active architecture that is characterized by the delicate balance of structural forces into stable equilibria through bundled configurations. The evolving material system combines Bending-Active and Tension-Active components. Material testing and physical form-finding studies provide a framework for predictive spring-based computation simulation under development by Sean Ahlquist in the processing environment. The series of experimental prototypes use GFRP Rods and polycarbonate connectors to explore the syntax and detailing of closed-loop bundled configurations. A computational overlay has been developed using Grasshopper to predict cross-sectional area based upon the local bending-radius of the prototype.

美国宾夕法尼亚大学建筑系 / U Penn

香港夜总会 / A Nightclub for Hong Kong

通过加强或是改变现有的居住模式以及参与及使用方式，建筑可以导致文化革新。影响的产生大多从室内空间开始。室内塑造了人们的空间体验，因此更能激发出文化变革。我们发展的动态 / 生成技术能够得到变化率以及形体变化的加速度和密度，而每个学生根据自己的兴趣和审美带来形体的多样性。我们的目标是利用多样性产生无法预测的建筑效果，包括从部分到整体的组织方式，与众不同的室内特点，以及基于形态的连续性由内而外地设计建筑形体。

Architecture generates cultural change by intensifying and inflecting existing modes of inhabitation, participation, and use. The creation of affects are most clearly pursued by starting with interiors. The interior shapes experiences within space and hence has more potential to generate cultural change.We developed dynamical/generative techniques that derive qualitative rates of change as well as accumulations and densities of form while each student developed their own interest in and sensibility for variation in form. The goal was to use variation to produce unprecedented architectural effects including part to whole arrangements, distinct formal interior features, and with morphological continuity develop a building from inside out.

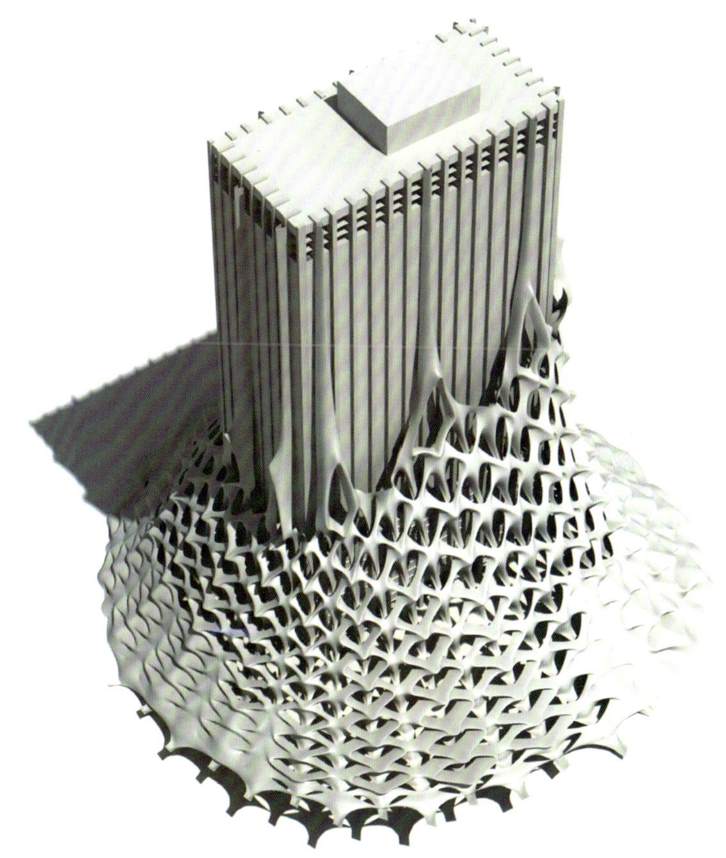

衬衫 / Skirts

本项目强调了海平面的上升及其对纽约市沿海部分的可能影响。使用参数化和多边形建模技术，我们开发了一套拓扑系统，它能够在现存建筑的周围包裹上具有保护性的大“衬衫”。这一部件具有可变的围合程度，能够根据建筑立面的水和光或者是水平面的高度进行个性化的设计。另外，薄膜也能够使得这个不断更新并部分沉没的城市可能创造出奇异的空间与程序效果，以及壮丽的氛围与体验。

This project addresses the rising sea levels and the anticipated impact they will have on the coastal parts of New York City. Using parametric and polygonal modeling techniques, we generated a system of topological components which can aggregate into large protective 'skirts' surrounding existing buildings. The components have various degrees of closure, which allows for individualized response towards either water or light based on the building elevation as well as the level of water. In addition the membranes also create novel spatial and programmatic conditions and speculate on the atmospheric and experiential qualities a retrofitted and partially sunk city could possibly generate.

塑料机器人系统 / Plastic Lotus Robotic System

本项目设想一个涌现的系统，利用计算和自然过程收集太平洋垃圾带中的塑料污染物，每个个体利用电流像水母一样运动、发现并收集塑料。个体也拥有自组织的能力形成更大的集群，甚至形成一个生态系统。该项目采用先进的3D建模技术，产生增量形态的变换，以达到广泛的适用性。个体被设计成具有面向特定目标的属性。机器人细胞能长成更大的组件和集群，同时也能被组织成建筑结构。

This project envisions an emergent system to collect plastic particle pollutants in the Great Pacific Garbage Patch using both computation and natural processes. Individual drones move with the currents like jellyfish to find and collect plastic. The drones have also the ability to self-organize into larger clusters and form an eco-system. The project uses advanced 3D modeling techniques to generate incremental morphological transformations in order to achieve a wide array of applicability. Individual units are being designed with specific goal oriented properties. The robotic cells can grow into larger assemblies and clusters as well as be organized towards the structure of a building.

美国南加州大学建筑学院 / USC

互动变体式建筑 / Alloplastic Architecture

互动变体式建筑希望探索研发一种能对人类运动产生反应的互动建筑的可能性。这个体系需要像人体一样有一种适应性的自由可变的结构。人体的这套结构包括抗压类（骨头），主动拉伸类（肌肉）和被动拉伸类（比如皮肤）。最终，一个表演艺术家可以和这个结构体系一起跳舞。不用身体接触，结构可以通过她的表现做出反应。用 Arduino 控制板和形态记忆合金 “弹簧”，Kinect 感应器可以追踪舞者的动态进而对整体结构进行再塑形。

Alloplastic Architecture explores the possibility of developing an interactive architecture that responds to human movement. The project involves an adaptive tensegrity structure that echoes the behavior of the human body which also operates as a form of tensegrity structure consisting of compressive members (bones), active tensile members (muscles) and passive tensile members (skin etc.). Eventually a performance artist dances with the structure that reacts to her presence without any actual physical contact. A Kinect motion sensor device tracks the movement of the dancer, and thereby reconfigures the entire structure through the use of an Arduino control board and Shape Memory Alloy 'springs'.

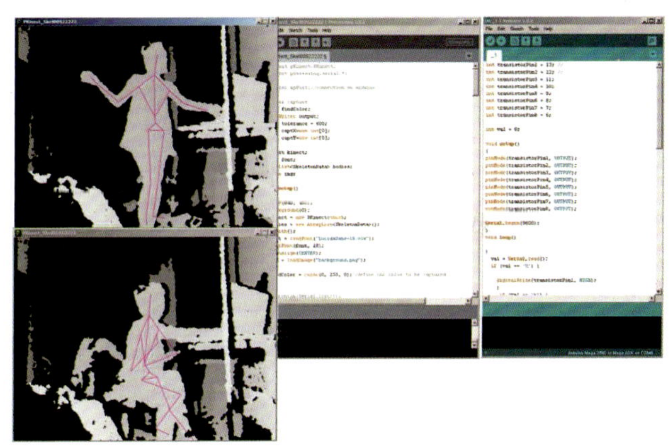

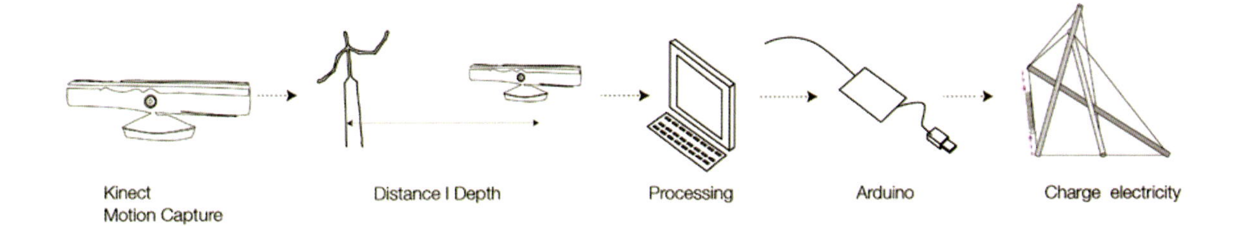
Kinect
Motion Capture
Distance I Depth
Processing
Arduino
Charge electricity

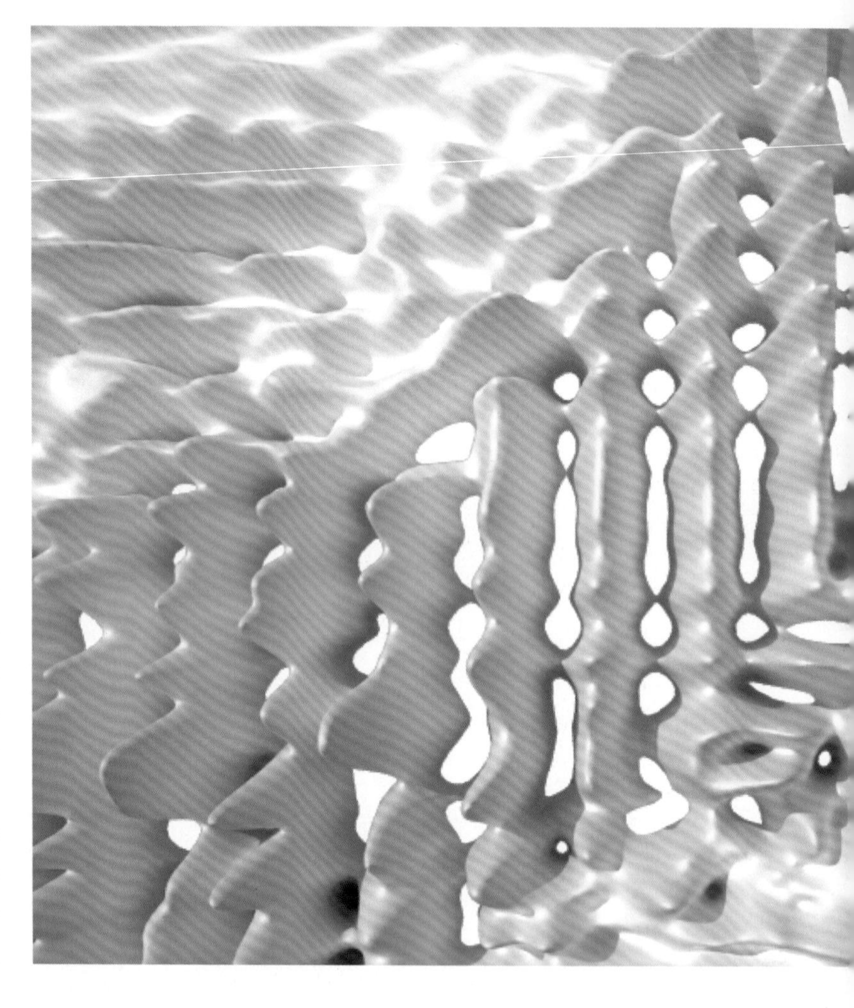

多方向的源字段 / Poly-Directional Source Fields

用 grasshopper 通过引入源点和曲线可以改变基础正弦波的振幅和波长。这个逻辑同样可以用在巨大尺度的几何当中，根据不同点和曲线的权重来控制震动的强度，从而产生线和点的循环。对于大尺度的装置来说，其制造过程从曾经的计算机数控粉末成型，发展成为现在在真空中用一种称为 PETG 的透明塑料成型。激光切割机根据不同的等高线进行切割，示例的颜色代表不同的几何特性，用乳胶将缝隙进行黏合。

Using Grasshopper, the amplitude and period length of the basic sine wave is changed, by introducing source points and curves. The same logic is then applied for the macro scale geometry, in order to generate circulation curves and points, with the strength in vibration based on the weight of different points and curves. Later for the micro scale installation, the fabricating process develops from CNC milling to PETG vacuum forming, with the laser cutter being used to create apertures based on different contour levels, then the color illustration is based on the geometry properties followed by the use of latex to cover the apertures.

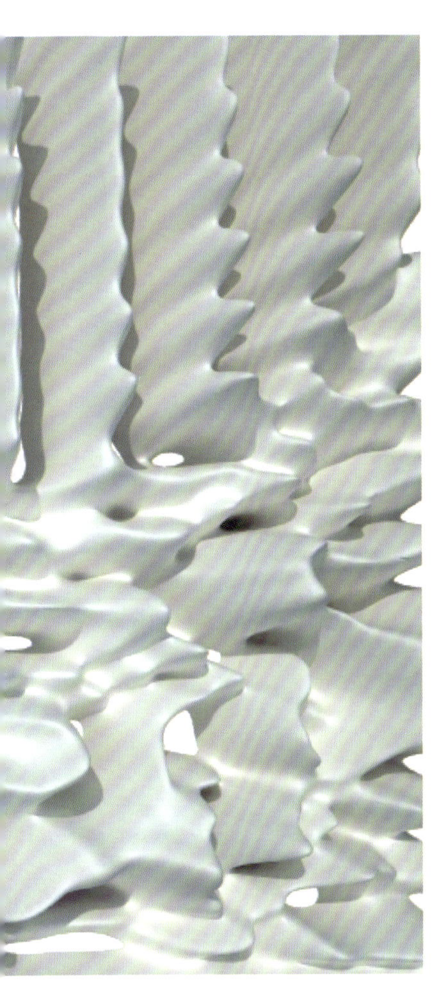

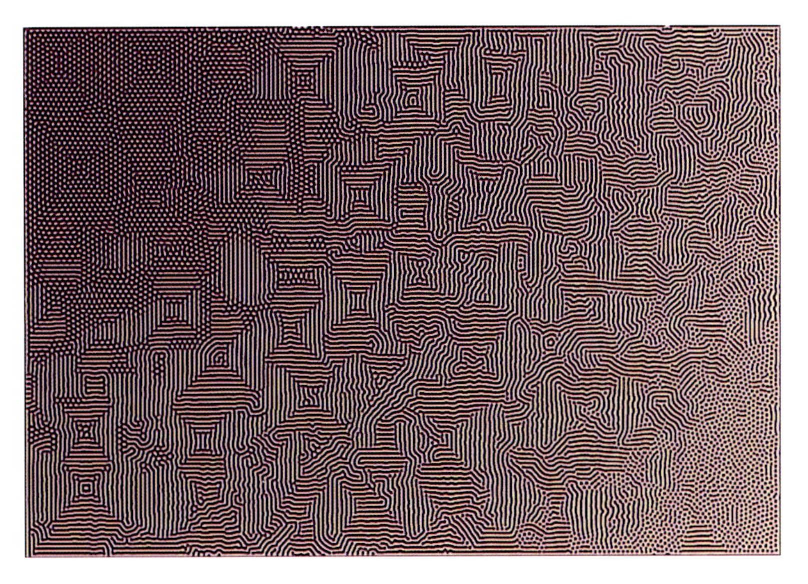

可控制的反应扩散系统 / Controllable Reaction Diffusion System

反应扩散系统是数学模型。它解释了在以下两个过程中一种或多种物质在空间变化中是如何分布的：物质相互生成的化学反应和引起物质沿空间表面铺开的扩散作用。在建筑设计中使用该系统有些困难。因此，系统的界限、核心及周边部分都需要根据空间中的作用更精确地定义。系统整体用Processing编写，可以产生点，进而生成网格面或表面。

Reaction - diffusion systems are mathematical models which explain how the concentration of one or more substances distributed in space changes under the influence of two processes: local chemical reactions in which the substances are transformed into each other, and diffusion which causes the substances to spread out over a surface in space. This system is difficult to use in architectural design. The boundary of the system therefore needed to be defined more precisely, along with the hollow and solid parts based on the functions of the spaces. The whole system is written in Processing which is used to generate points and then meshes or surfaces.

美国耶鲁大学建筑学院 / Yale University

展亭组装 / 'Assembly One' Pavilion

由学生设计、组装、建造的展亭是从一个角度看很坚实、很厚重，另一个角度看很轻盈，几乎完全渗透的组装展亭。它的结构交替隐藏和展现纽黑文夏季国际艺术和思想节的内容。展亭由薄铝板组装而成，在学校加工车间内的 CNC 机床上进行切割。展亭打开两面形成通风和通透的感觉，同时视线聚焦在节日的主舞台上。在参观者经过展亭时，1000 多块板创造了不断变化的反射和色彩效果，为节日创造了迷人的核心空间。

The 'Assembly One' Pavilion, designed, fabricated and erected by seminar students, is solid and massive from one angle, lightweight and almost entirely porous from another. The structure alternately hides and reveals its contents, the information center for New Haven's summer International Festival of Arts and Ideas. Constructed from thin aluminum sheets CNC cut in the school's fabrication shop, the pavilion opens up on two sides for ventilation and transparency, focusing views toward the festival's main stage. Over 1000 panels create shifting effects of reflection and color as visitors move around the pavilion, creating an engaging heart for the festival.

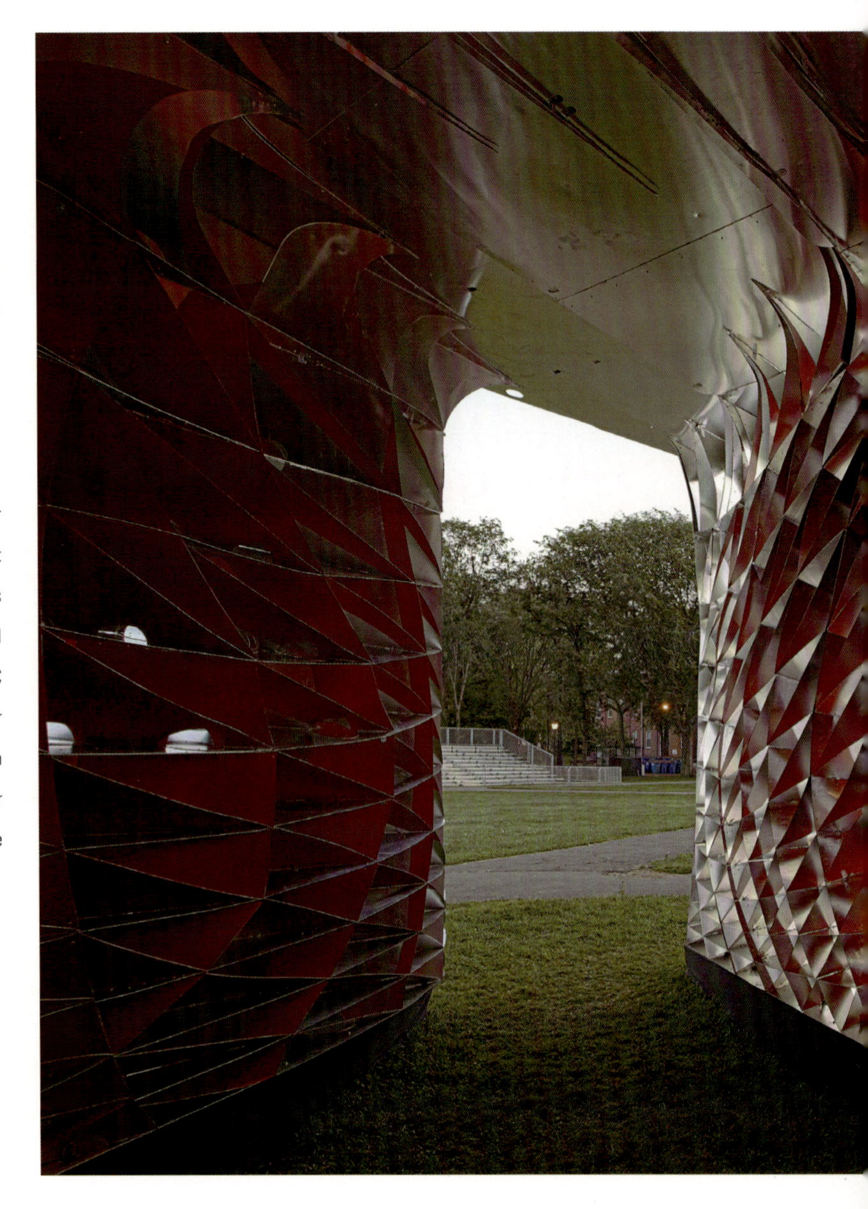

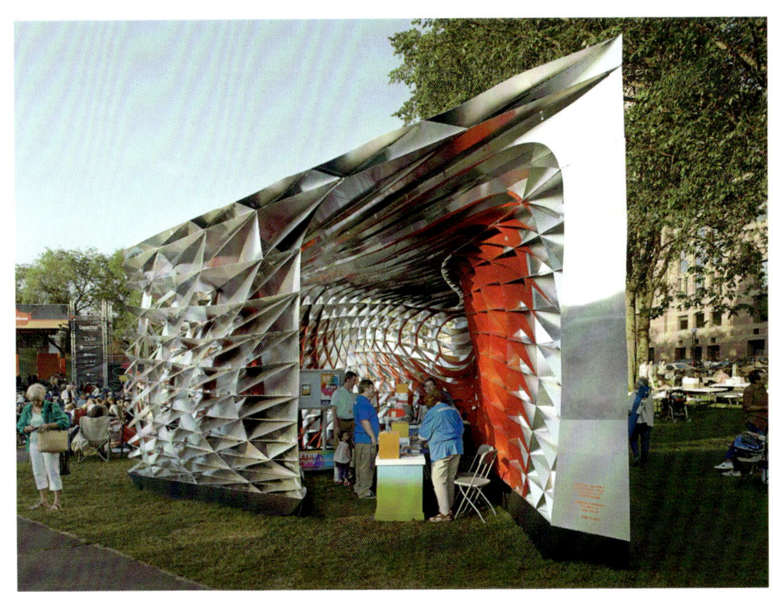

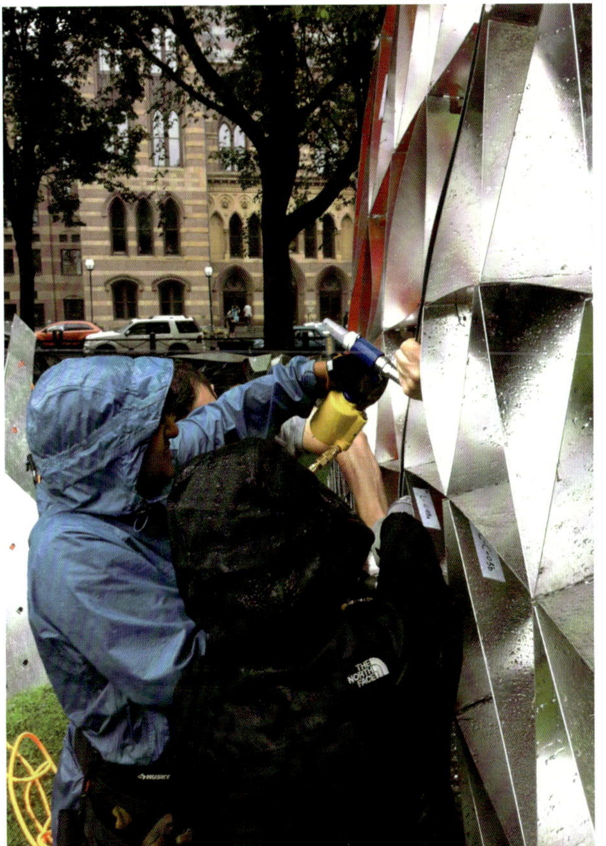
THE NORTH FACE

凌乱的几何 / Disheveled Geometries

“粗面光边石工”被用于描述建筑的最极端的表面纹理种类。粗面光边石工现在应用得更广，这种技术用于使建筑变得不光滑。随着参数化、计算化和分形方法提高了研究小尺度高分辨率物体的能力，这次研讨会重新涉及粗面光边石工这一主题。建筑师可设计非常现代的纹理，它可服务于如通风、遮阳、隔热、防水、支撑、发电、物理防御或纯粹的美学等各个方面。学生们学习历史上粗面光边石工这一工艺，并且以此为基础设计建造一个大尺度的原型。

The term 'rustication' is used to describe architecture's most extreme category of surface textures. Rustication now takes more effort rather than less, and skill is measured in moving away from architectural smoothness instead of toward it. With the ability to parametrically, algorithmically, and fractally manage matter at increasingly small scales of resolution, this seminar revisits the topic of rustication, where architects design unapologetically contemporary textures that might act in the service of everything from wind dispersal, shading, insulation, water shedding, grip, power generation, physical defense, or pure aesthetic effect. Students study methods of rustication throughout history and use this research as a foundation to design and produce large-scale prototypes.

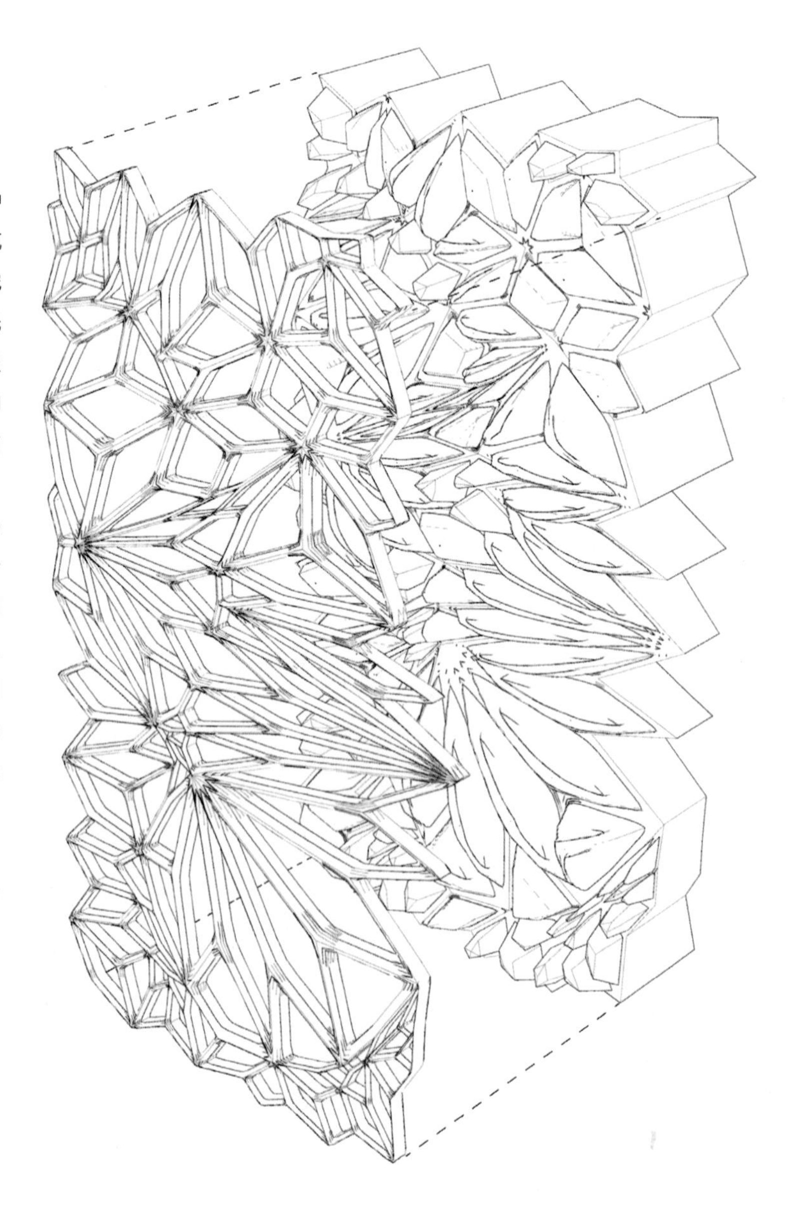

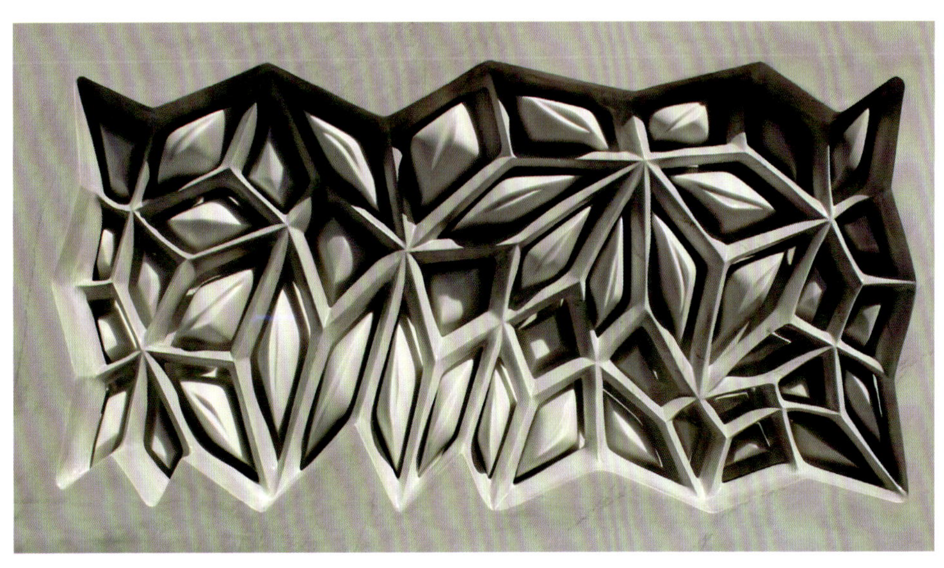

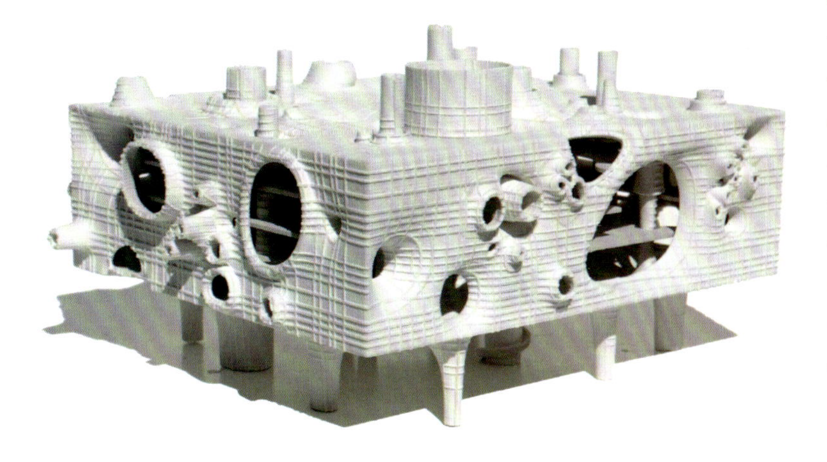

复杂的编织 / Wired Integrated Composite Knit

本项目将创新性的材料工艺和技术传送装置相结合，创造了自支持及表现性的空间。这种技术可以通过一个可变的编织方式适用于服务、结构和完整的围墙。当提供了插入式的基础设施时，表皮可产生不同的体量变化。当独立或阵列遍布在空间、地板、吊顶以及垂直连接构件里，并且连线充电，它就能够提供电力、无线网、一般工作照明和辐射冷却功能。在它全部的配置中，智能系统是单独传递的，通过编织布局，用户的参与使其动起来。

This project combines innovative material processes with a technology delivery apparatus to create self-supporting, expressive spaces. This technology can accommodate services, structure and full enclosure through a flexible knit construction. The surface's repertoire yields different volumetric possibilities while providing plug-in infrastructure. Standing alone or arrayed throughout an existing space, the floor, ceiling and vertical connections become wired, charged surfaces that deliver power supply, wireless internet, general and task lighting and radiant cooling. In all its configurations, the intelligence of the system is built into a single thread, distributed through the weave and made active through user engagement.

英国建筑联盟建筑学院 / AA

集群化工作 / Swarm Works

该项目是一个使用集群机器人所形成的原型自治系统，作为结构和空间环境的代理在松散填充环境中进行测试，形成了本项目的主要设计核心。机器人被设想为基于群体阶层筑巢行为规则的社会性昆虫，自行组织形成一个类似于白蚁生态的人造社会生态系统。该系统的生成始于对环境条件参数的评估，将其上传到机器人的记忆库后，最终将其作为集群包在目标区空投进行部署。

The project is a proto-typical autonomous system employing the use of swarm robotics, which forms the crux of our proposal as it is tested in loose fill environments as an agent of structure and space. Conceived as social insects based upon the rules of stigmergic social hierarchies, the robots organise themselves as an artificial social ecosystem that can be likened to that of termite ecologies. The deployment of the system begins with an assessment of environmental conditions and parameters, which are then, uploaded into the memory banks of the robots, which are then deployed as clustered packs in target zones via air drop.

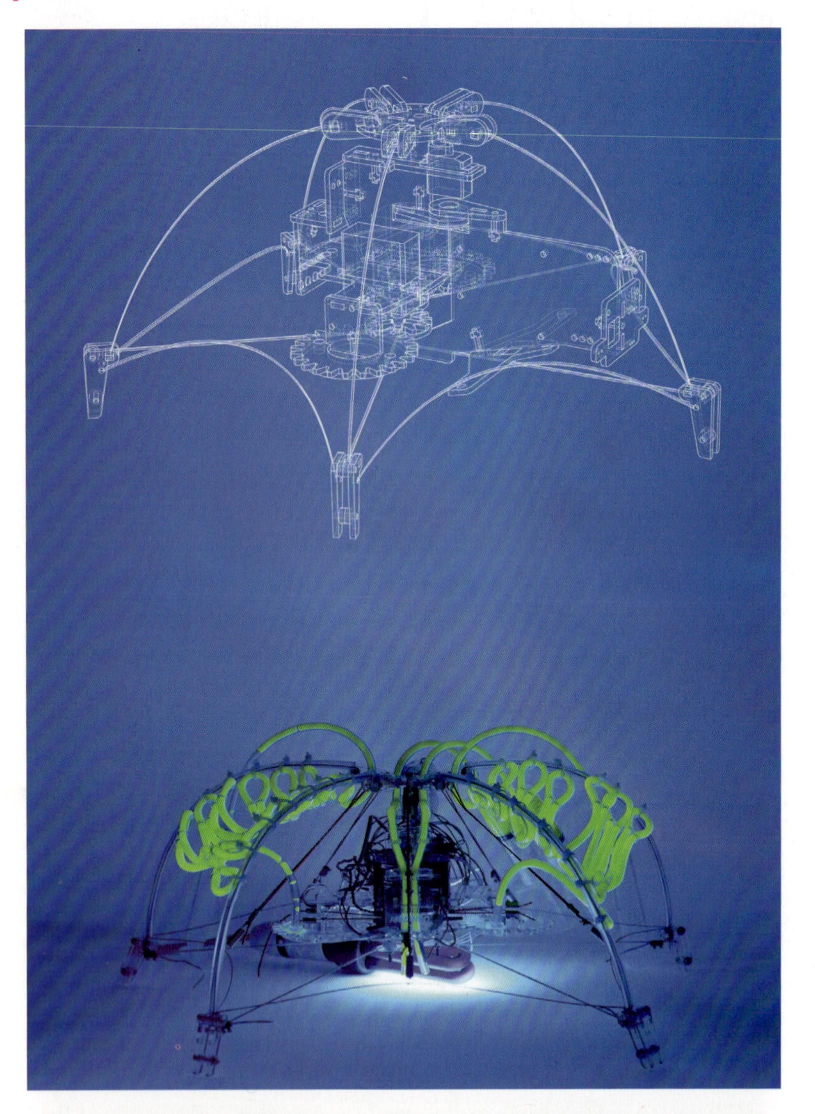

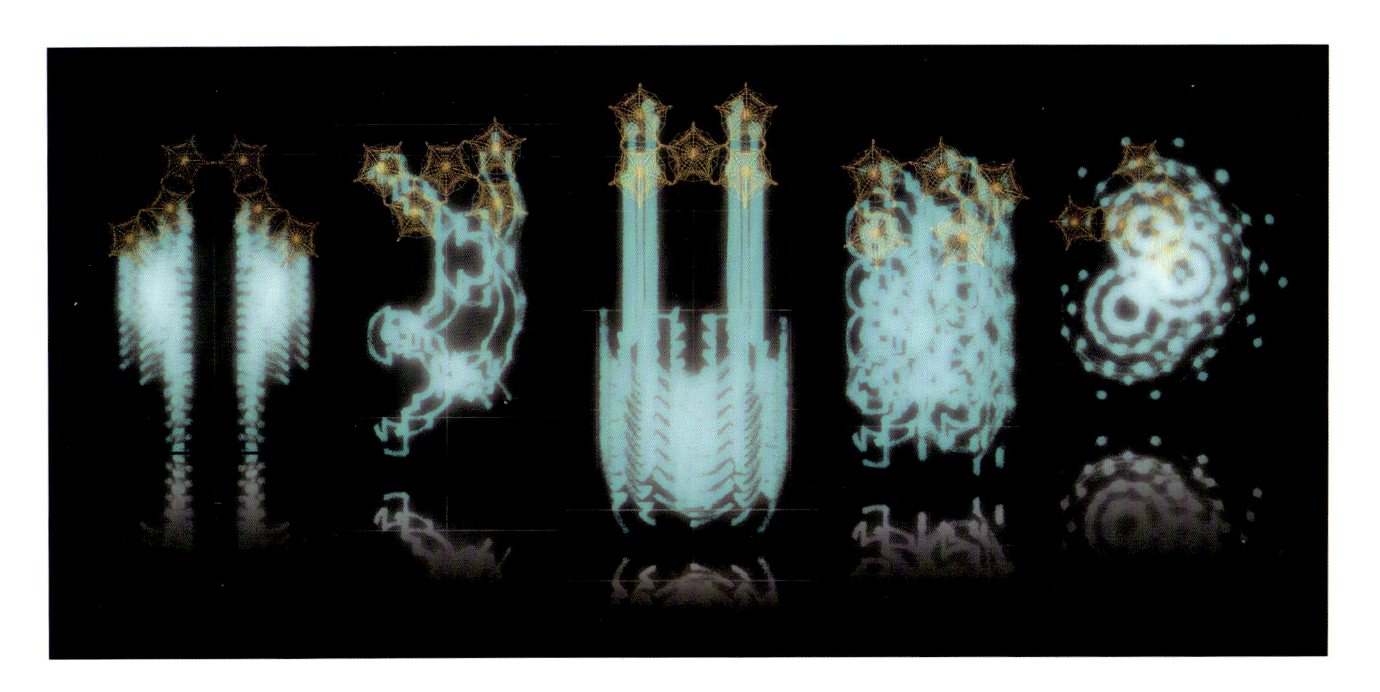

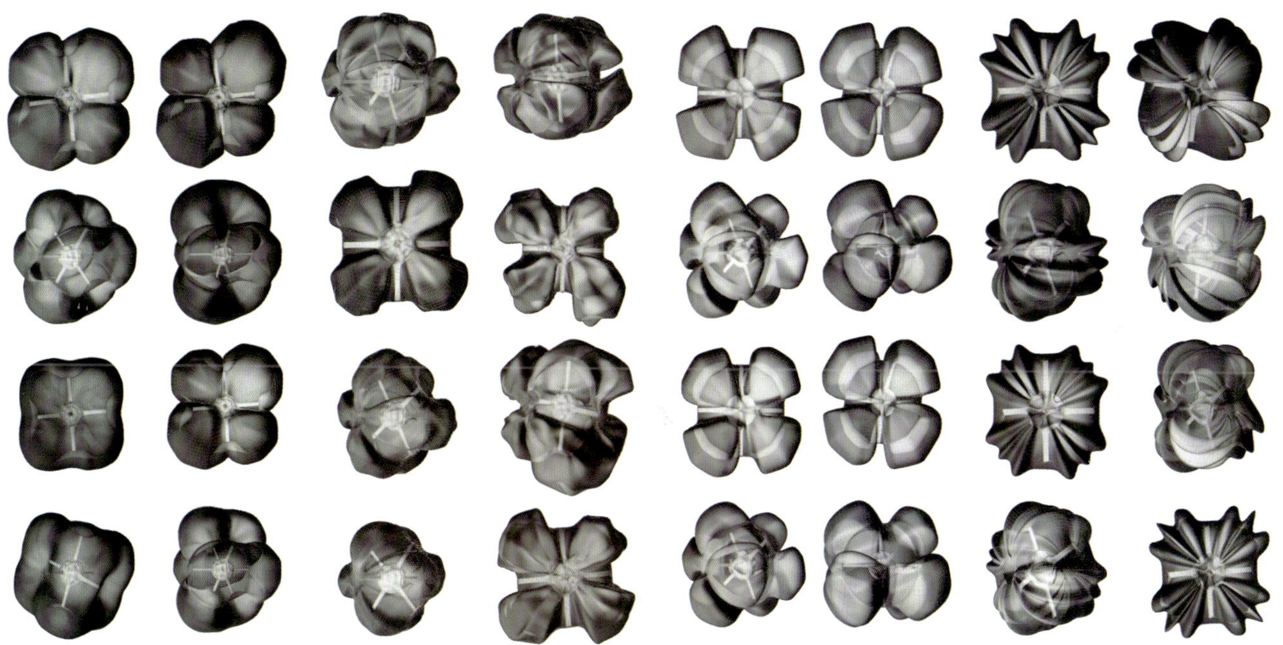

活水技术 / Life Aquatech

本项目研究的是作为室内和室外媒介的建筑系统之间的关系，以及从空气系统转移到水性体系的建筑设计体系。通过把流体的行为作为生成设计的方法和功能标准评估工具，该项目建议构筑一种利用“水”作为建筑技术统和元素的建筑系统。通过水的收集、存储和分配，这个建筑原型旨在创造一个有凝聚力的建筑环境，通过不同的以水为基础的建筑系统之间的相互作用，达到设计美学和建筑性能的完美融合。

This project investigates the relationship between building systems that mediate between interior and exterior and architectural design by shifting from air-based systems to water-based systems. By focusing on the behaviour of fluid as part of both generative design methodologies and evaluation tools for functional criteria, the project proposes the deployment of a building system where water plays an integral role in the building tectonic. Through the collection, storage and distribution of water, the building prototype aims to create a cohesive architectural environment through the interaction of different water-based building systems, resulting in a fusion of design aesthetic and building performance.

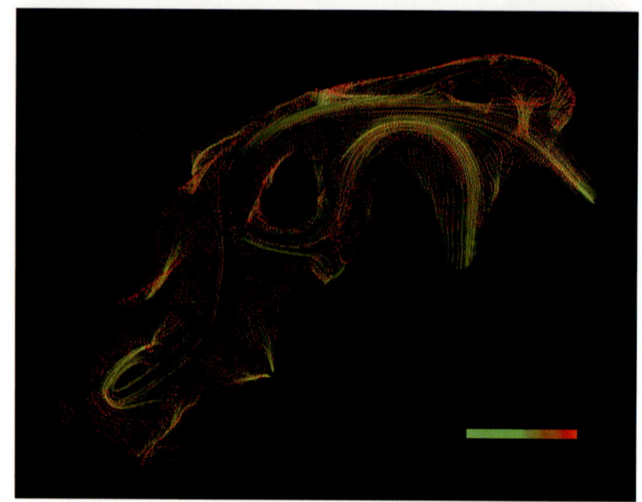

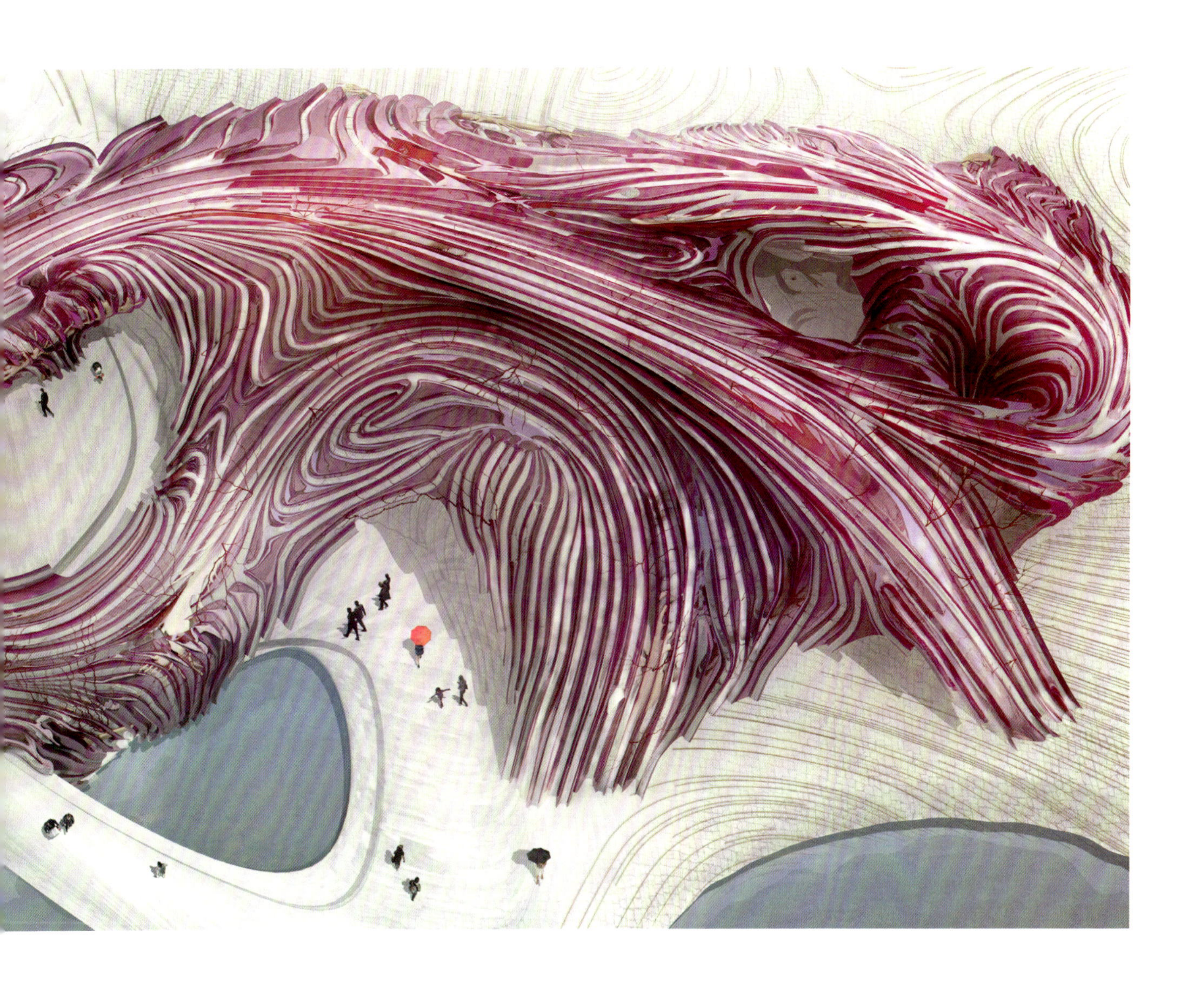

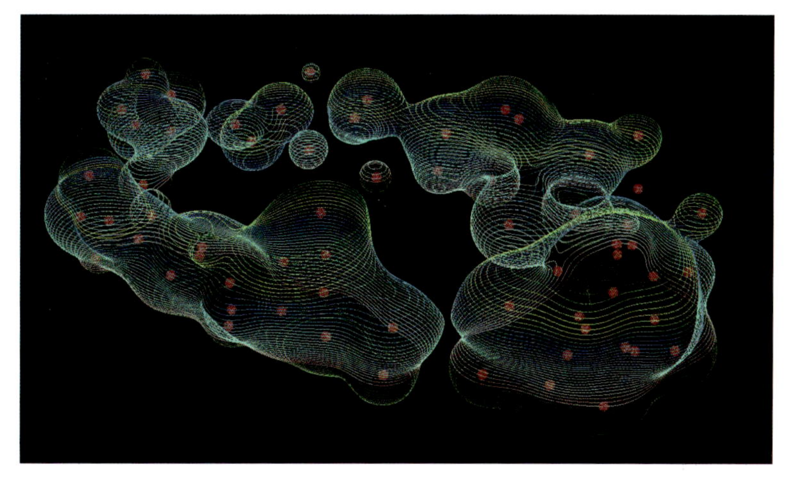

改革 / Reformation

我们开发了一个数字化互动设计系统用于挑战传统的建筑设计方法。建筑从人与人的互动开始发展，而不是根据视觉组合来构筑建筑的几何形状。这与平面的美观无关。该项目实际上是个建筑机器，在设计环节中直接取代了建筑师的角色。它以一种前所未有的方式集成了人类的行为，允许利用其集体行为来定义建筑设计过程中的关键环节。设计师的作用变成仅仅是一个对过程进行部分限定的代理。

We have developed an interactive and digital design system that challenges the conventional methods of architectural design. The building develops from human interaction, instead of the visual composition of the geometrical shapes of the building. The beauty of a plan is irrelevant. The project is in fact an architectural machine, which takes the hands of the architect directly out of the design process. It integrates human behaviour in an unprecedented way and allows its collective behaviour to define key aspects of the building design process. The role of the designer turns merely into an agent for a process that partially defines itself.

奥地利维也纳工艺美术学院 / Angewandte

二元循环 / Binary Cycles

汉尼·让希指导的该设计是通过发展一个概念的、实际的、具有批判性的技术生成新的、有趣的、具有前瞻性的建筑。我们鼓励学生考虑所有有关空间的因素，不仅是集合、形式的手段，还有经验、实验的结果。设计通过物理模型，模拟计算机生成图像、数据、图解等手段着重强调“概念的证明”。这一概念也在装置设计中有所体现。

Hani Rashid's studio program focuses on the development of conceptual, practical and critical skills for the making of new, compelling and forward thinking architecture. Students are encouraged to consider all aspects of spatial form making in terms of not only geometric and formal approaches but also as experiential and experimental outcomes. The studio places a great deal of emphasis on a 'proof of concept' approach utilizing physical models, animation and computer-generated imaging as well as diagrams and data sets. This also covers installation work in support of all spatial and architectural arguments, design concepts and strategies.

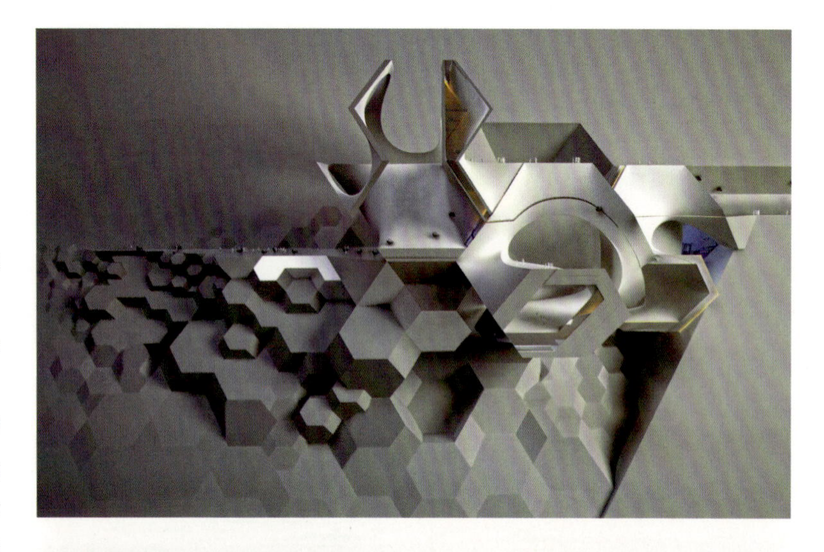

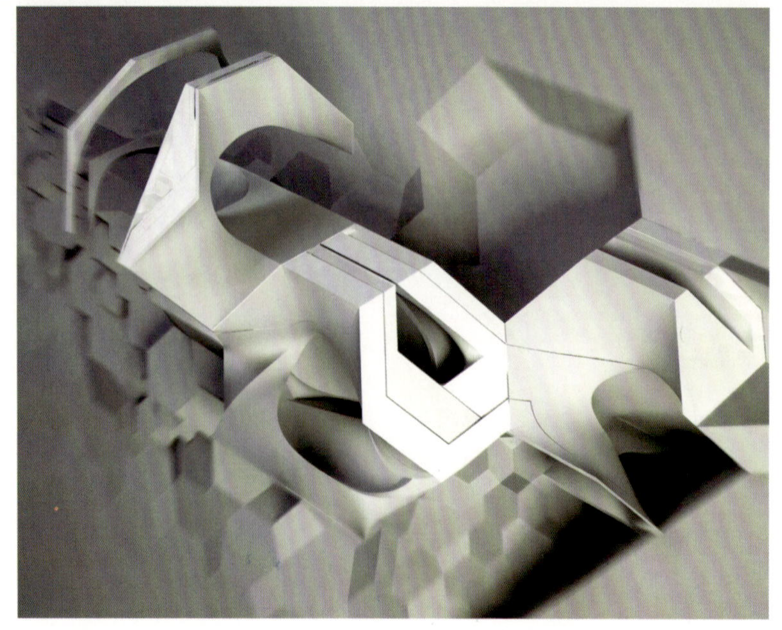

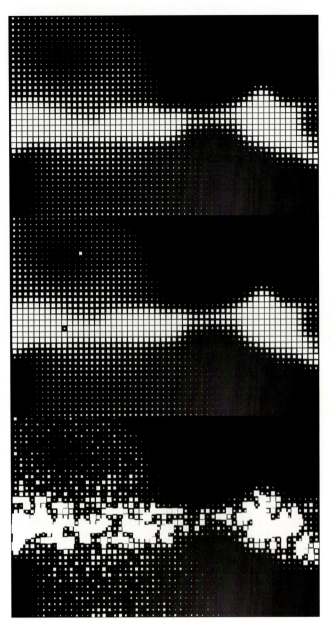

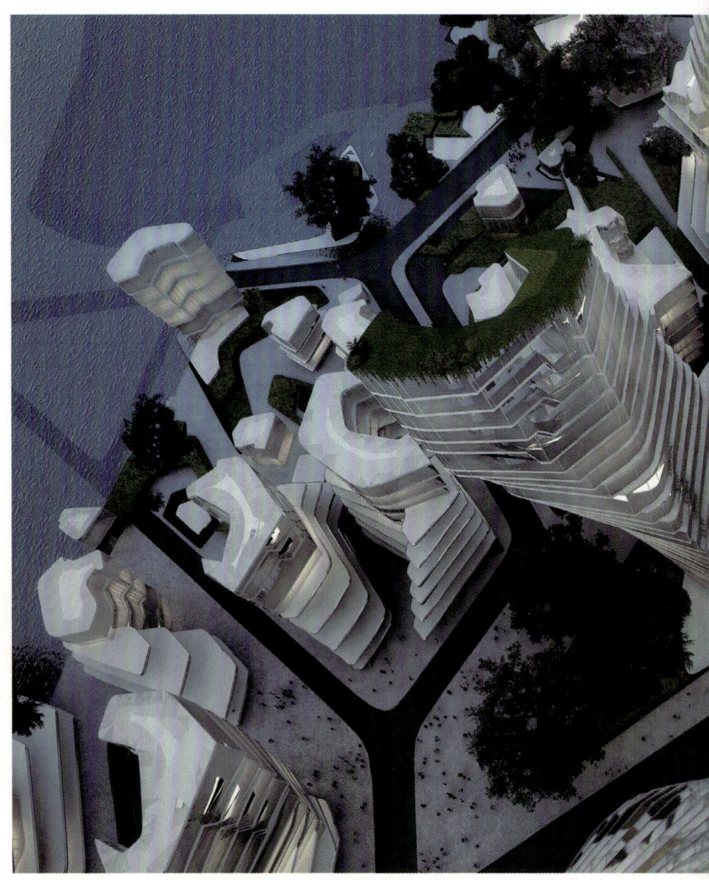

秩序 + 复杂 + 矛盾 / Order + Complexity + Contradiction

参数化设计的力量在于秩序与复杂。但是这两种参数的综合导致了过度超定的结果。更多参数将会改进、提高这个复杂的系统，但也会持续提升决定论的程度。唯一能够将这个系统从过度决定论中解放出来的参数类型是“反参数化”，即将场地矛盾的概念重新引入参数化城市中。本项目探究了参数化主义，以及如何使参数化城市设计成为一个秩序、复杂、矛盾的综合作用系统。

The power of parametric design lies in its capabilities of correlation (order) and differentiation (complexity). However, the sum of these two types of parameters leads to an overdetermined result. The introduction of even more parameters will advance the system in its sophistication, but will also constantly increase the degree of determinism. The only type of parameter that is capable of liberating the system from the dogma of overdeterminism is a so called "anti - parameter", that re - introduces the notion of local contradiction into parametric urbanism. This project investigates within the studio's agenda of 'Parametricism', how parametric urban design can become the synthesis of order, complexity and contradiction.

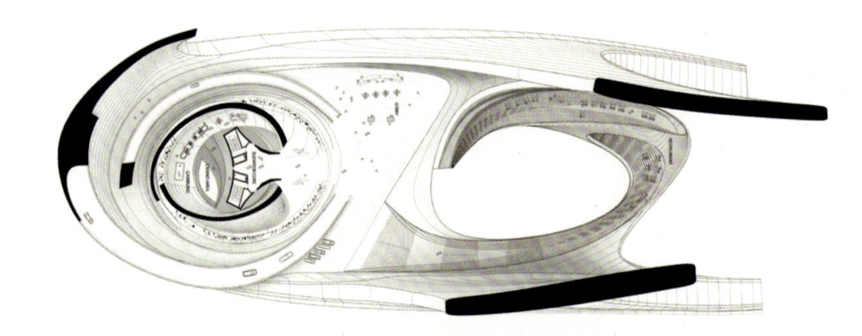

垂直长带 / Vertical Strip

本设计是关于不透明材料表面的互动，其特点是一些轻型易碎结构。通过这种结构形成的建筑从垂直方向上形成一种空间序列。设计的题目是一个赌场，但并不同于拉斯维加斯的那种赌场。设计场地位于临近拉斯维加斯的内华达内，在胡佛大堤和包帕大桥之间。这个休闲度假场所包括多种 21 世纪功能，例如娱乐、旅行演奏、混合武术搏击、赌博以及奢华生活。

This project is about the interplay of vast opaque surfaces that incorporate poché and lightweight, fragile structure. The environments developed by these distinct architectural languages are exploited and distributed vertically to create a variety of extreme spatial sequences. The project is a design for a casino resort not dissimilar to those in Las Vegas. The project is located on a dramatic site in Nevada close to Las Vegas between the Hoover Dam and Bypass Bridge. The resort caters for a range of various twenty-first century vices such as entertainment - including concert venues, and Mixed Martial Arts fighting - gambling and luxury living.

英国伦敦大学巴特利建筑学院 / Bartlett

合成可构造性 / Robofoam

Robofoam 是一个关注由泡沫材料产生干扰互动的项目。这种材料具有算法和机器产生的非线性高精度特点。整个项目建立在一种回馈循环机制上。这种机制在数字设计与行为模拟间建立联系，应用到了多代理系统、材料研究以及利用机器臂和泡沫枪的建造过程。同时，重新定义材料和设计方法也能够发掘其美学与结构性能，创造一种有弹性的建筑，从而产生一种新的景观。

Robofoam is a project that looks at the interplay of noise sourced from the material behaviours of foam, characterised by a high degree of non-linearity, simultaneous to extreme precision of algorithmic and robotic production. The project synthesis is developed through a constant feedback loop between digital design and behavioural simulation using Multi-Agent Systems, material research and fabrication processes with an industrial robotic arm and modified foam gun. At the same time both redefining material and design methodologies the project is able to mine aesthetic and structural capacities enabling an evolving resilient fabric of architecture which generates new synthetic landscapes.

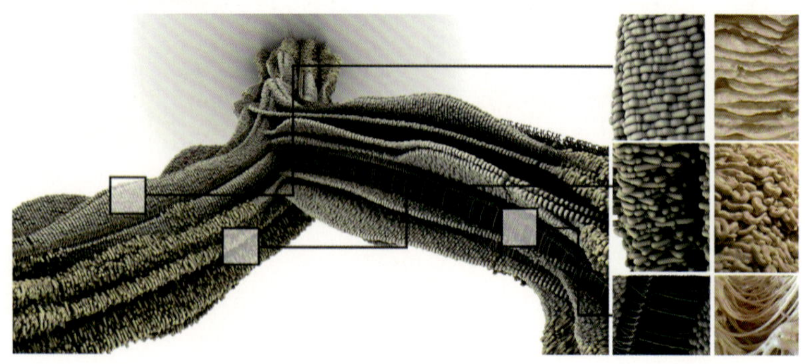

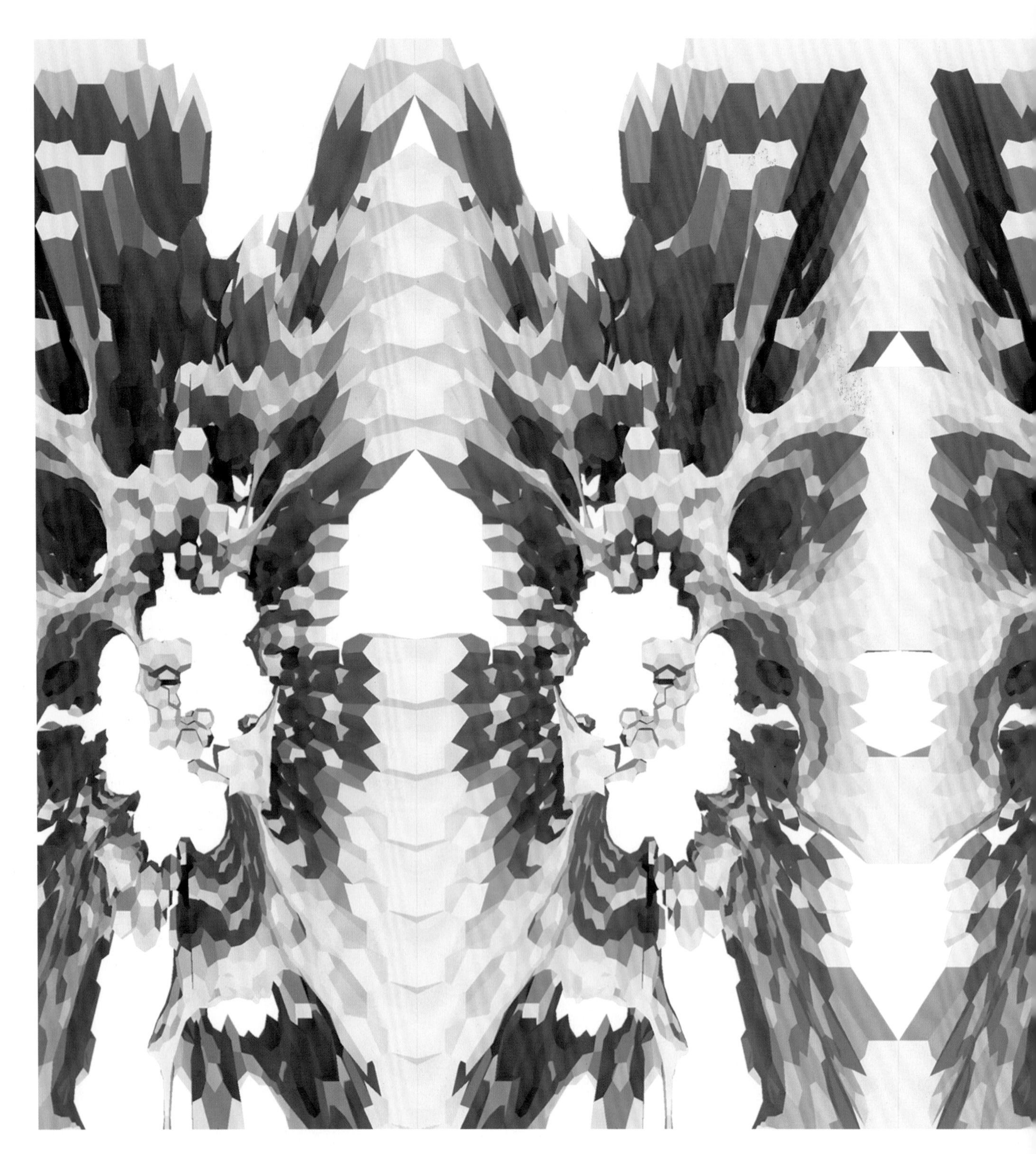

深度材质 /
Deep Texture

深度材质应用于现阶段复合材料概念的研究以及对大尺度建筑的补充。它试图重新理解复合材料的应用以及在一种材料系统中多种条件下的无缝转变。在一个大尺度的项目中，重新基于人体与自然思考建筑构造与类型，在人体中，各种不同功能与性质的骨骼、肌肉、血管、神经、皮肤组织等多系统组合成一个附属的完整的功能体。

Deep Texture deals with the current disconnect between recent material research as a concept of multi materiality and its implementation into architecture on a macro scale. The aim is to reinterpret the use of synthetic composite materials and to develop seamless transitions of varying material conditions within one material system. On a macro scale the project initiates a rethinking of architectural tectonics and typology in reference to nature and the human body where bone structure, muscle, vein, nerve and skin tissue, multiple systems of varying performance and material properties, become one functioning entity of dependencies.

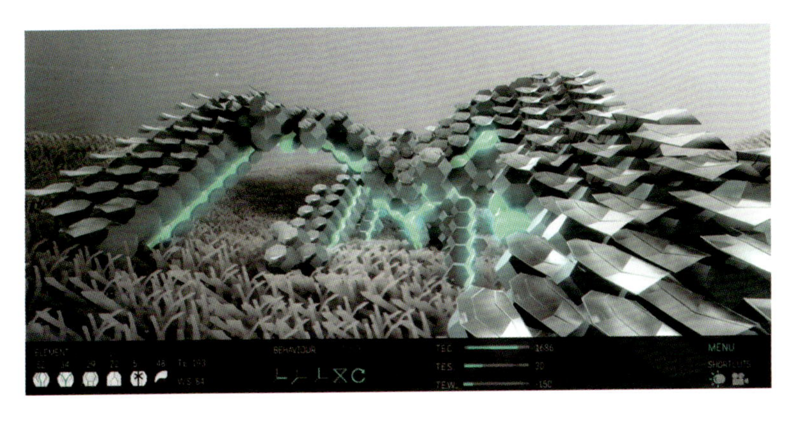

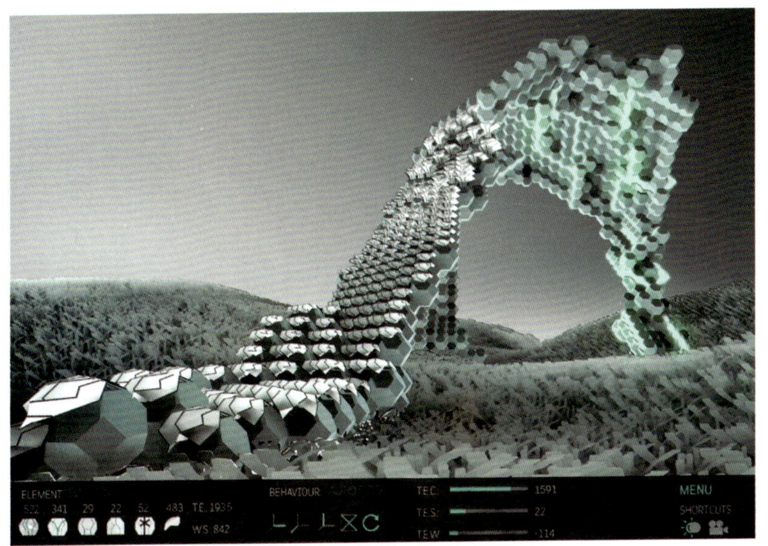

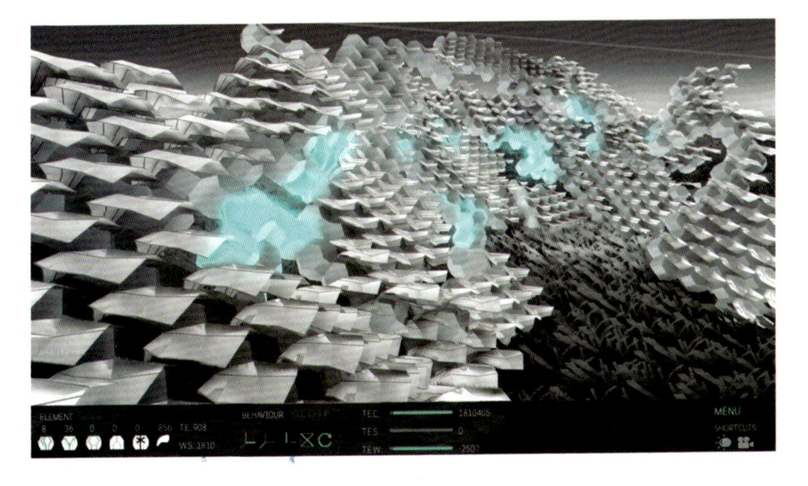

星火虫项目 / Wireflies Project

星火虫项目探寻一种通过玩的设计过程生成一个建筑的可能性。使用者通过决策探求新的设计策略。项目的概念是创造一个敞开的沙箱游戏，有一个离散的可以被组合、拆解的工具，可以产生不同的聚合效果。这个游戏探究一种新的砌块，能够收集周围的风能。使用者能通过光的图案来重新定义建造过程中的电路和信息网的类型。

Wireflies project explores the potential of creating an architectural system through a playful design process. It is about the user's creative expression, who through decision making explores new strategies in architectural design. The idea is to create an open sandbox video game as a discrete kit of parts that could be assembled and dis-assembled resulting on different aggregations. The Wireflies game explores new building blogs of architecture working around the collection of the wind energy and its distribution. Players can use light patterns to intuitively redefine the topology of the circuits and information networks that run through the building fabric.

丹麦皇家美术研究院信息技术与建筑中心 / CITA

避风港 / Haven

该项目提出在哥德堡的岛上建立一个温泉疗养院。空间框架结构像轻盈的面纱一样设立在敏感的景观上。最初的研究包括在空间框架中的材料策略、无形的空间细分、变异和统一空间的影响。研究应用了基于物理的薄膜张力模拟和菲涅尔透镜的光学作用。透镜有双重作用，形成双重的空间框架外壳部件。独立的几何体由特定位置的结构、光接收面的距离和预期效果等参数计算得来。

The project proposes a spa and hospice on the island of Saltholmen. A space-frame construction is developed as a lightweight veil sitting on the sensitive landscape. Initial studies develop strategies of material and immaterial spatial sub-division, differentiation and effect within the unified space of the space-frame. The strategies are explored using physics-based simulation of tensile membranes and the optical effects of Fresnel lenses. The lenses perform a dual function, doubling as elements of enclosure for the space-frame. Their individual geometries are computationally determined by interrogation of parameters pertaining to attributes of specific siting within the structure, focal distance to light receiving planes and desired effect.

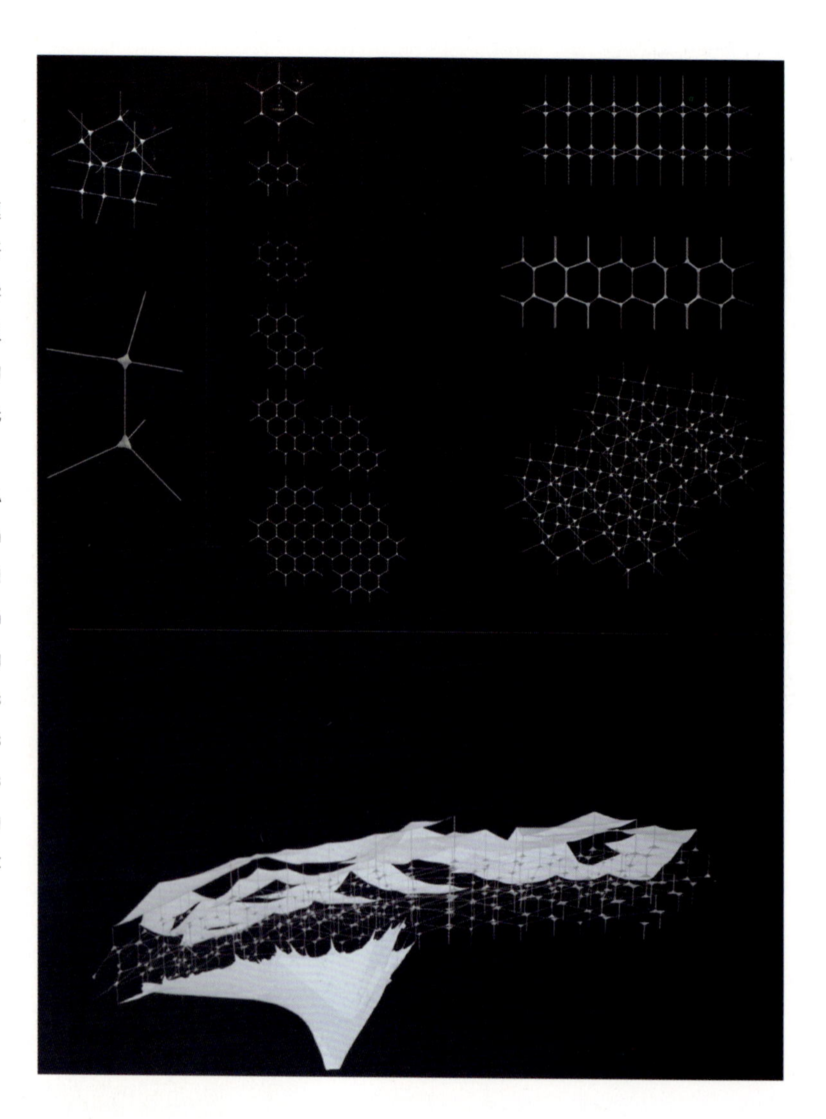

哥德堡的纪念塔 / Monument on Saltholmen

该项目考虑了在机器人和材料科学领域的新兴技术如何应对气候变化带来的挑战。在哥德堡的岛上将设立一座 50 年后将淹没在上升水平面下的纪念碑。它的石灰石基地、雨水回收、持续的风能是一个机器人社会的基础。为了提炼原料形成三维打印结构，形成连续的计算环与风对抗，创造的声学效果会让我们记起小岛，同时新的纪念塔为人们提供了避难所。

The project speculates on how emerging technologies in robotics and material science can establish ecologies to handle the challenges of climate change. Here a monument is proposed for the island of Saltholmen, which will disappear under the raising water-levels in 50 years. Its limestone base, collected rain water and the energy from the continuous wind are the foundations for a society of robots. The seinteract in order to refine the raw matter into 3d printed structures, shaped in a constant computational loop to engage with the wind. The created acoustic effects will remind us of the island, while the new monument provides a sanctuary for humans.

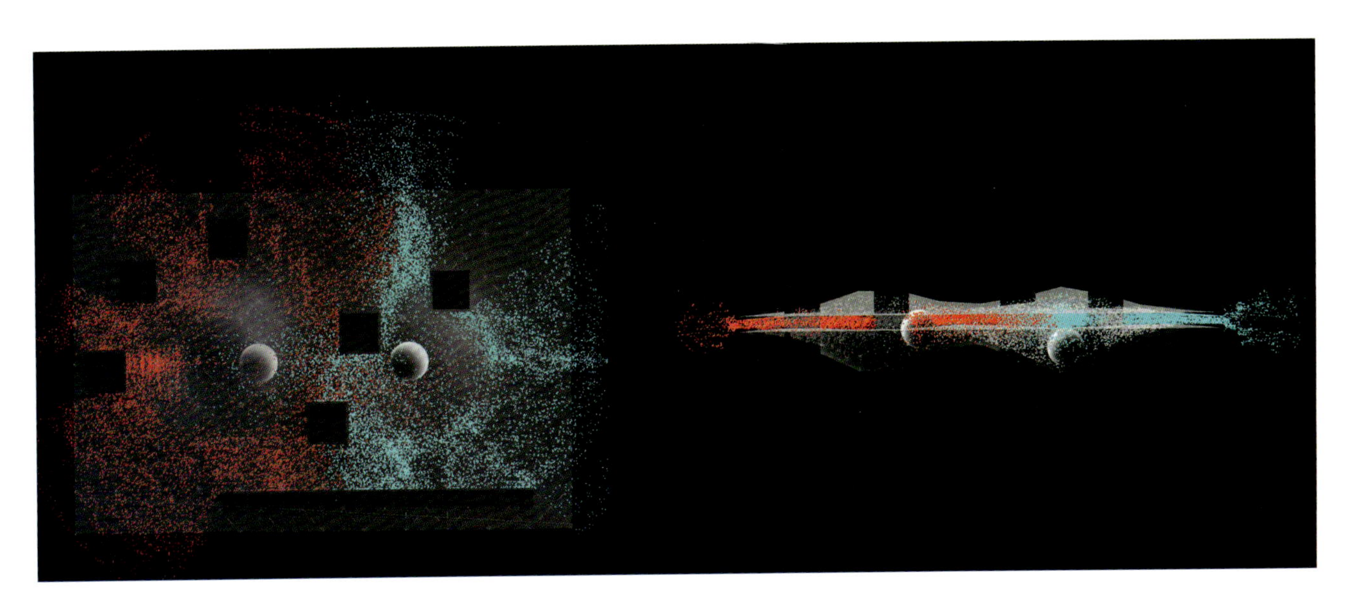

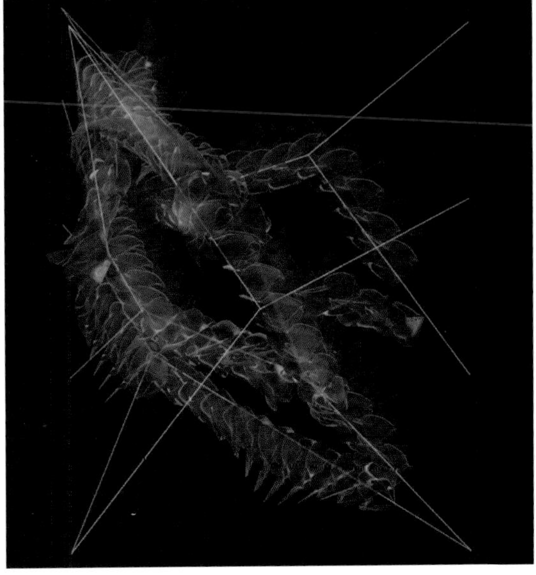

光的瀑布 / Cascade of Luminescence

该项目探讨了在计算设计实践应用中物质文化的延伸。它的概念从轻盈出发，并对空间框架和模块化结构特别关注。数字运用形成对系统、模块以及相关设计概念和技术的认识。通过检测材料交点的数字信号，学生们提出在亲自动手制作过程中描述现实、图解与细节之间的重要转译关系。本项目最终由建构思索变成新的具有潜力的数字设计、数字模拟和数字建造方法。

This project explores the extension of material culture that occurs with the application of computational design practice. It takes its point of departure in notions of lightness, with a special focus on space-frames and modular structures. Digital investigations generated an understanding of systems, modules and associated design concepts and techniques. Examining the digital through its intersection with the material allows students to query the underlying translations between description and matter, schema and detail through a hands-on engagement with the processes of making. The project results in a resolved architectural-tectonic speculation into the new potentials for digital design, digital simulation and digital fabrication.

荷兰代尔夫特工业大学 / TU Delft

Agriflux 项目 / Agriflux

Agriflux 项目的研究主题是调查数字信息与建筑之间的空间关系。这一生成性的系统主要包括以下三个连续的范围：代理、结构和行为。系统中有两种代理，一种具有较高等级，另一种具有较低等级。它们被进一步细分并基于代理组成结构，它们可能成为基础设施、功能性组件或者只是依据系统的思想逻辑在分层的网络定义中循环往复的数据流。这些代理通过行为本能地进一步拼装，而这些行为本能地让它们具有展开、被差异化的场地吸引或排斥以及其他多代理系统的运动。

The project "Agriflux" is a thesis research which aims to investigate spatializing the relationship between digital information and architecture. The generative system takes three contingency areas into account: agency, structure and behavior. There are two sets of agencies, one being the higher order and second being the lower order. These are further sub classified and structured under agents which could either be infrastructural, functional or even just routing streams from layered network definitions based on system thinking logic. These agents are further embedded with behavioral instincts which helps them to deploy, be attracted or repelled by differentiated spots and other agent swarm movements.

巴黎食物中心 / A Food Hub for Paris

本项目的目标是解决巴黎市区的食物生产与运输问题。为了让项目的计划能够与现行的动态都市系统相关联，我们使用了不同的计算技术。细节实时模拟（在 Processing 平台中开发）和受集群行为准则控制的多代理系统是在环境中获得最优功能空间组织的主要工具。从模拟结果中衍生出来的数据用来形成一种形态过程，这一过程通过使用参数化方法和环境软件限定出三维的空间结构。

The project aims to deal with the problem of food production and transportation in the city of Paris. In order to integrate the project's proposal with the existing dynamic metropolitan system different computational techniques were used. In detail real time simulation (developed in Processing) and multi-agents systems guided by the swarm behavior principles were the main tools used to define an optimal functional spatial organization within the surrounding. The data derived from the simulation were employed to inform a morphological process that through the use of parametric and environmental software define a 3-dimensional spatial structure.

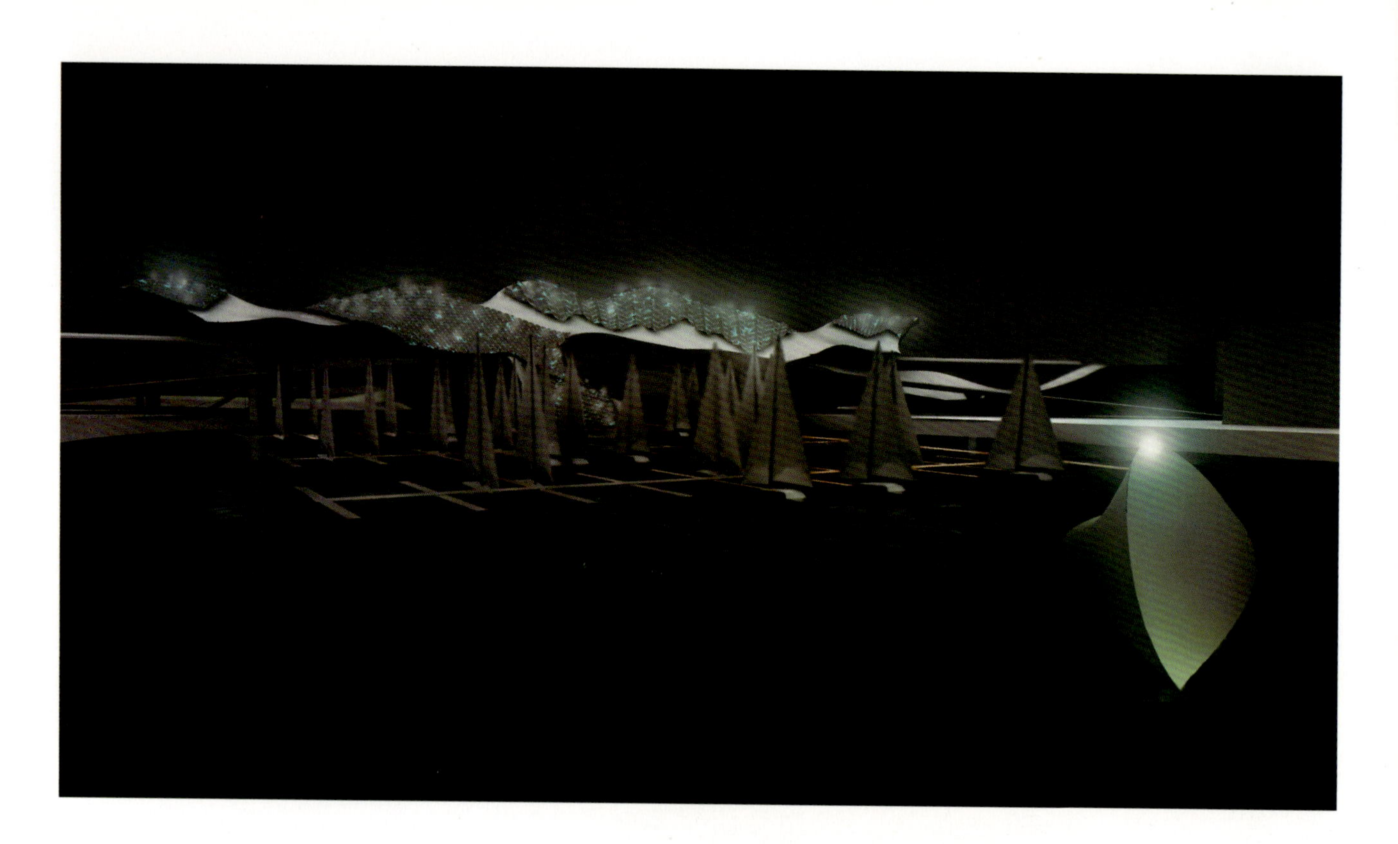

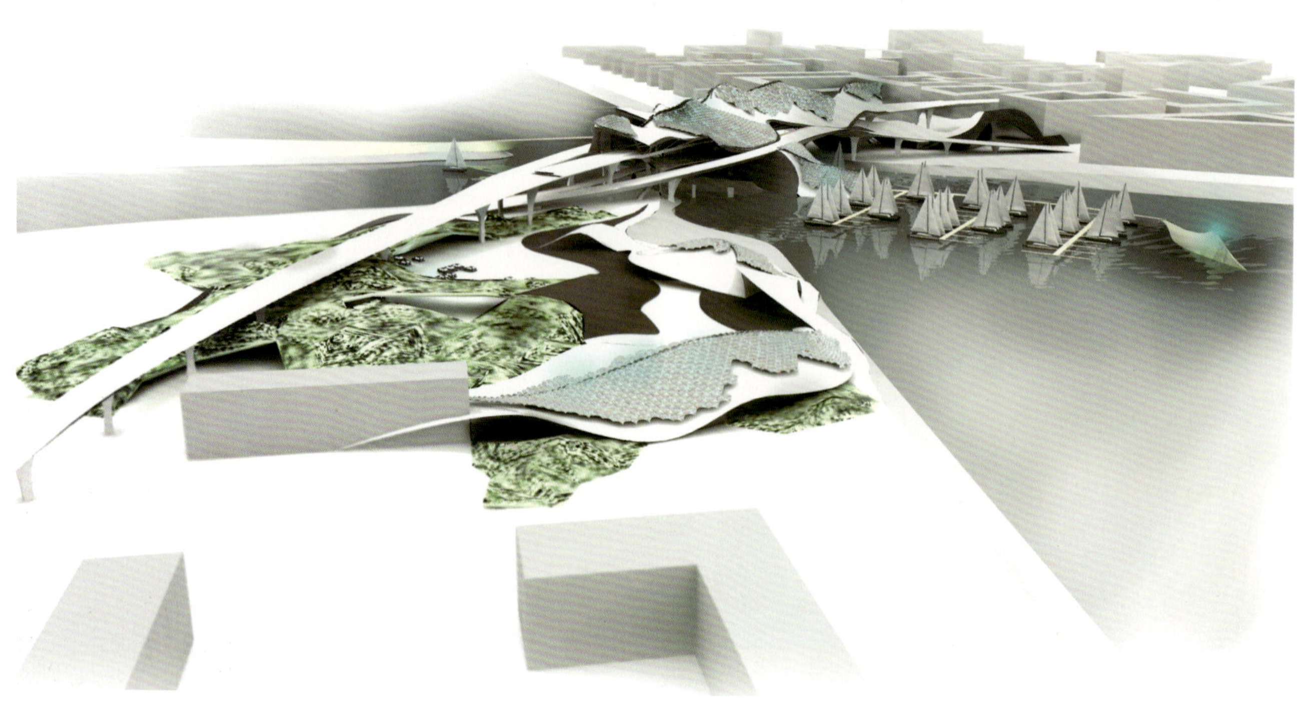

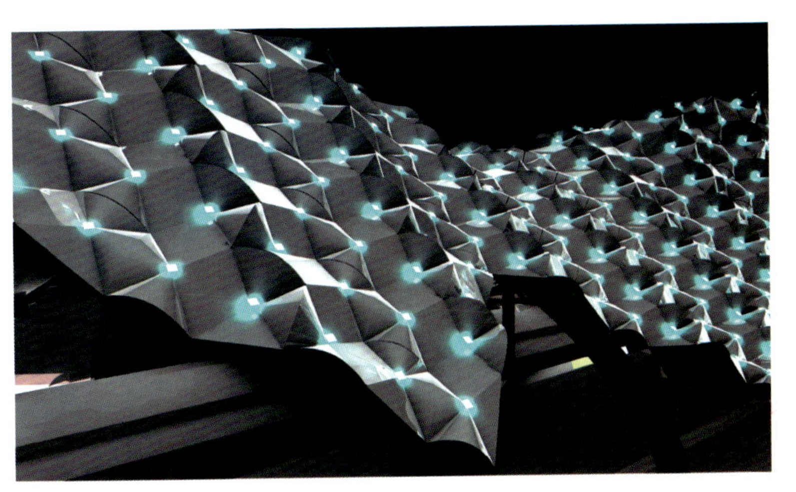

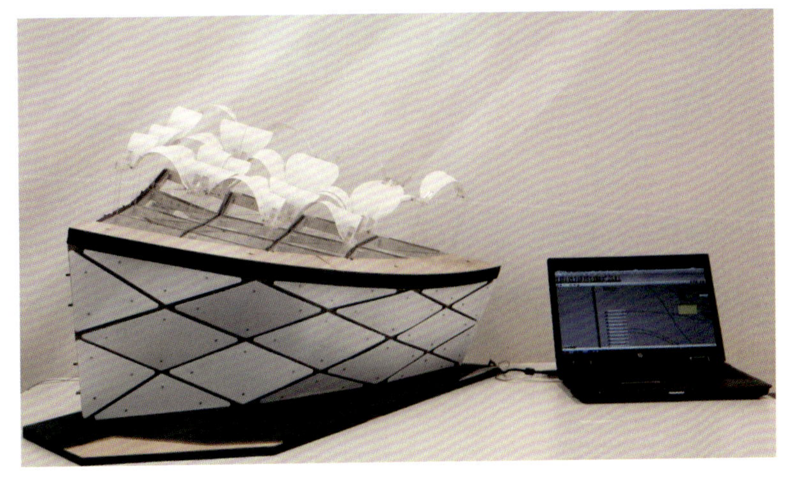

Transferium Almere 2.0 项目 / Transferium Almere 2.0

本项目的内容是一种多形态的交换中心，它是使用网络化城市方法动态总图工具（DMPT）和与风相关的可变建筑几何体进行开发的。DMPT 能够计算出新的城市拓展开发中的变化。这一工具基于自下而上的集群行为，从而能够实时实现城市地块的设计与功能变化，通过反馈环路重新定义总图与建筑之间的关系。风的行为是基于压力差的，这是其基本数学准则。通过 CFD 模拟我们能够根据 DMPT 获得的变化形态生成出潜在的能量网络，进而使用遗传算法进行优化。

This project for a multi-modal transfer hub, was developed using the networked urban solution Dynamic Master Plan Tool (DMPT) and wind-informed geometry of Transferium buildings. The DMPT computes changes in the development of new city extensions. The tool is based on bottom up swarm behavior that accommodates design and functional changes of urban blocks in real time, redefining relationships between master plan and building through feedback loops. The mathematical principle of the wind operates on pressure difference. Based on CFD simulations a potential power network is generated on proto-shape of the Transferium obtained from the DMPT, further optimized with genetic algorithms.

德国德绍建筑学院 / DIA

DUNElab 项目 / DUNElab

DUNElab 项目是基于自动拓扑探测方法，对沙丘的烧结建构。这一过程完全依靠脚本算法，无论是地形生成还是生成形式。项目是一个沙漠景观的总图规划，这个景观设计通过集群机器人扫描，确定出相应的几何条件，以便在场地中生成网格结构。这是一个非相加的建筑设计过程，它通过含有多层复杂空间的人工地形与景观进行深度的互动。

DUNElab is a project that proposes sintered tectonic constructions in the sand dunes of the desert based on automated topological detections. This is a fully scripted project that is entirely dependent on algorithms to both evaluate the site and generate the forms. The project is a masterplan informed by the desert landscape which had been previously scanned by swarming robots in order to classify certain geological conditions, so as to eventually produce the network structures on site. It is a non-additive architectural approach that is highly interrelated with the landscape through the creation of an artificial terrain that has multi-layered spaces and mechanic ornamentations.

抗载形态学 / Load Reactive Morphogenesis

这个项目以骨骼行为为主要原则，用以定义材料受力系统，并应用到适应性建筑设计的结构中。基于 Wolf 法则——描述应对变化压力的骨质内材料的变化，建筑的结构系统利用拓扑关系根据不同荷载优化材料密度。利用可以动的单元实现不断完善、进化的结构系统，它们能够识别自己的位置以及应对自身的应力。通过在所有砖块中嵌入处理器单元，系统能够根据环境做出改变，也可以针对使用者做出改变。

This project uses bone - tissue behaviour as a principle for the definition of a material distribution system into adaptable structures for architecture. Based on Wolf's law, which describes the transformations of material patterns in bones in response to changing stress distributions, the building system uses topological optimisation to optimize material densities according to variable loads. To achieve a continuously evolving structure the system requires the introduction of movable units, able to sense their own location and the forces acting upon them. By embedding simple processor units in all bricks, the system can respond to environmental changes, caused either by users of the building or environment changes.

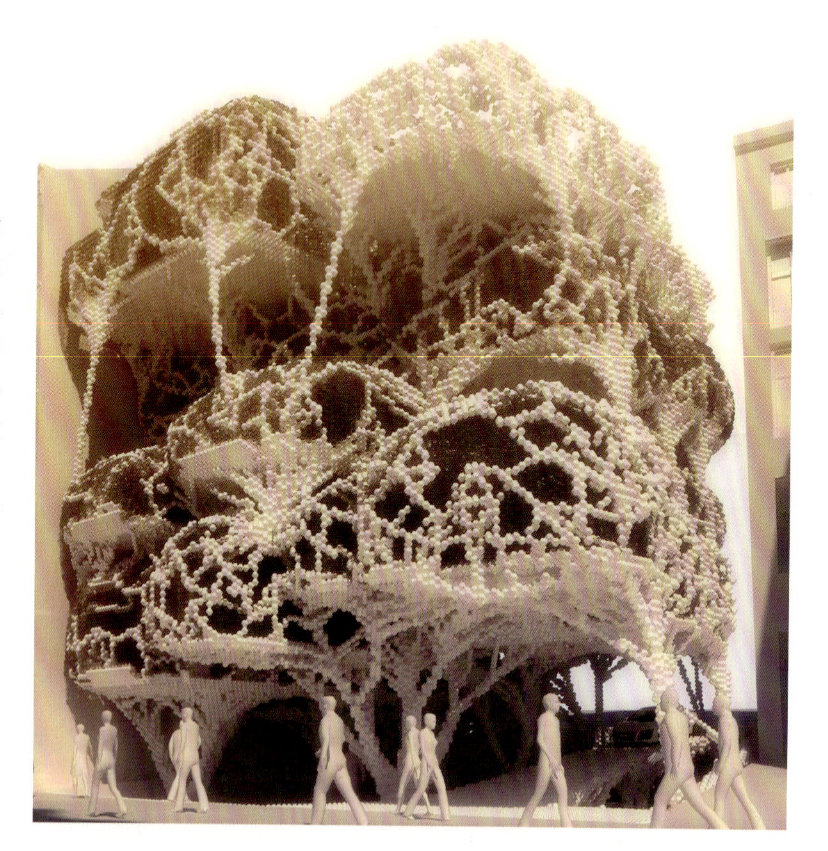

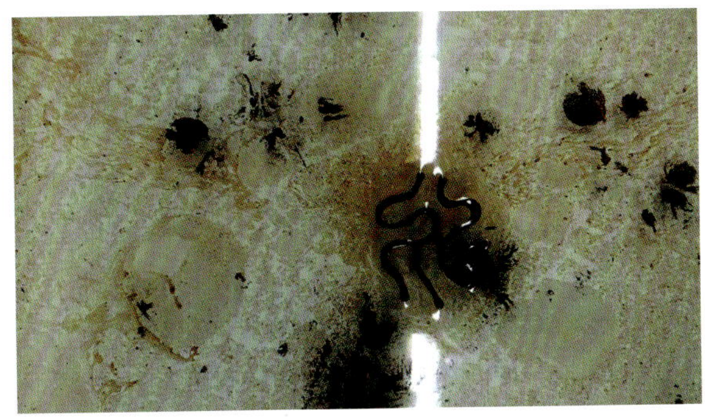

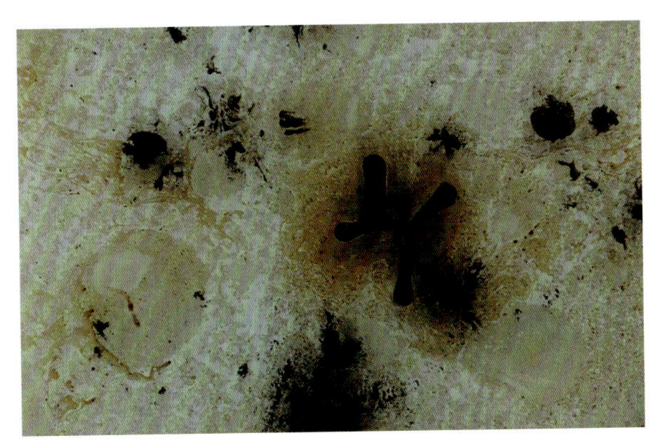

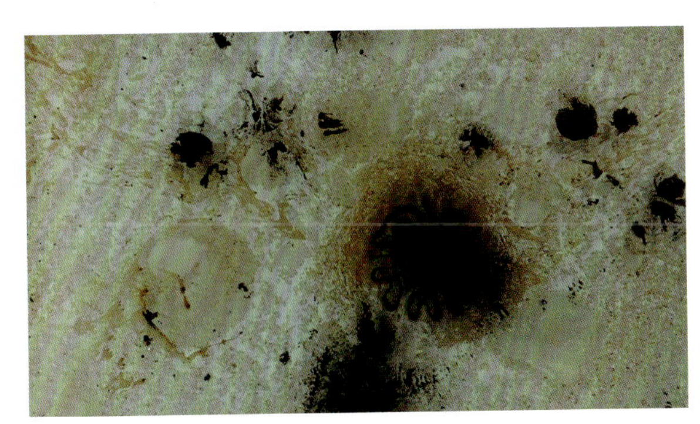

铁质流体 /
Ferro Fluids

这个项目探讨了在互动环境中铁质流体的应用潜能。在不同磁场吸引下进行了多组实验，并且利用 Kinect 追踪装置研究基于 processing 的干扰。最终，在两块水平玻璃上放置铁质流体，并在下面用小车带动一个可以移动的磁场，以控制铁质流体形成的形式。小车的移动是根据 Kinect 使用者的身体活动决定的。

This project explores the potential use of ferro-fluids within an interactive environment. Several tests were performed examining the behavior of ferro-fluids using various types of magnetic attractions in various configurations, and the use of Kinect motion tracking device was also researched using a Processing based interface. Eventually it was decided to contain the ferro fluid within two horizontal sheets of glass above an enclosure where battery powered vehicles supporting magnets were able to move around and control the patterning of the ferro fluid above. The vehicles can be controlled interactively by the body movements of the user through the use of the Kinect device.

西班牙加泰罗尼亚高级建筑研究学院 / IAAC

多层纤维 / Stigmergic Fibers

多层纤维展现了在各种环境和材料条件影响下纤维材料的发展前景。这个课题基于对纤维和生物的研究。我们在不同的生长条件下进行多种实验从而获得植物生长和结构的信息。通过喷射独立纤维来创造可控制、可重复的聚集模式。加入黏性液体，增加纤维之间的表面张力，产生更长的纤维连接表面。这样，该项目可以不断叠加，生成一个短暂的、融自然现象与机器建造为一体的建筑学意义上的栖息地。

Stigmergic Fibers tackles the prospect of fiber aggregation under the influence of varying environmental and material conditions. The project was initiated by research into plant fiber and biology. Various experiments were conducted on differentiated growing conditions to extract information on the behavioral methodologies of plant growth and structure. Separate individual fibers were sprayed to create conditioned and repeatable aggregation patterns. A water based adhesive was added to increase surface tension between the fibers and produce longer spanning fibrous connections and surfaces, allowing the project to be to scaled up to produce an ephemeral architectural habitat that would merge natural phenomena with robotic fabrication.

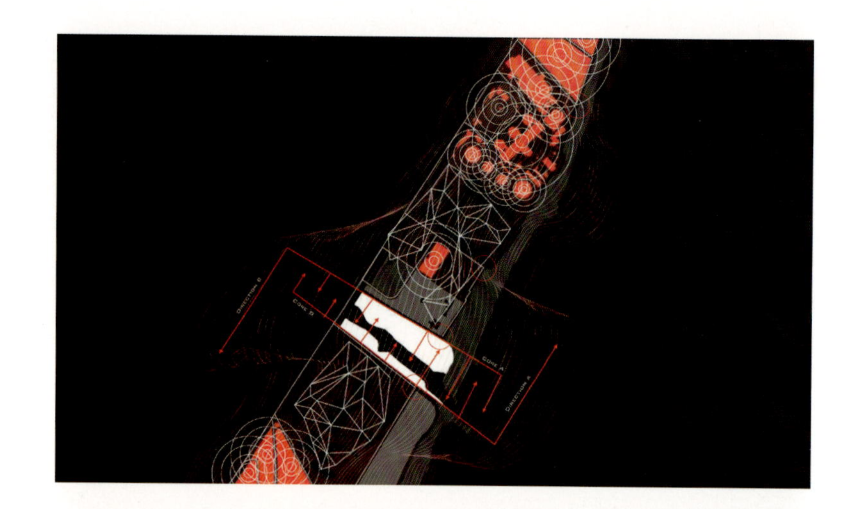

贝索斯河的互动跨越 / Interactive Spanning of the Besòs

贝索斯河穿过巴塞罗那，注入地中海。它分割了城市，遍布该区域的众多基础设施元素加重了这种现象。本设计希望重建这个已经被分割的城市组织，以数字技术为基础提出策略：扫描早期形成的现状，以细胞再生的有机行为为原型建立多维联系，从而建设一片新的区域。通过动画模拟、进一步探索涌现的城市肌理，以提高可读性，强调关于城市景观潜能的理论。

The Besòs River crosses the entire city of Barcelona pouring into the Mediterranean Sea, strongly dividing the urban fabric, which is further exacerbated by multiple infrastructural elements crossing this territory. This intervention aims to reconstruct this disconnected urban tissue, by developing a strategy based on digital techniques: scanning pre-existing territories and constructing a new territory through multidimensional connections based on the organic behaviour of cell regeneration. The emergent urban fabric is further explored through animated simulation, in order to increase legibility and to radicalize the discourse around urban scale interventions, in terms of their scenographic potential.

3D 抗重力打印 /
3D Anti-Gravity Object Modeling

这个题目建立了一种新的加法式建造方法。可以让要创建的 3D 物体的每一个表面都足够光滑，且不需额外的支撑结构。用这种创新的方式，我们可以在打印过程中忽略重力的作用。此方法利用的是 3D 切割，而不是 2D 分层。2D 分层不能充分体现物体的结构，而 3D 切割能准确地遵从定义形状的受力情况。总而言之，这种打印方式能够生成几乎所有尺寸和形状的物体。

This project proposes a new additive manufacturing method that allows 3D objects to be created on any given surface independent of its inclination and smoothness, and without a need of additional support structures. By using this innovative way of extrusion it is possible to neutralize the effect of gravity during the course of the printing process. This method creates 3D curves instead of 2D layers. Unlike 2D layers that are ignorant to the structure of the object, the 3D curves can follow exact stress lines of a custom shape. Finally, this printing method can help manufacture structures of almost any size and shape.

奥地利因斯布鲁克大学 / Innsbruck

集聚大厅 / Aggregated Lobbies

该项目是面向城市的“集聚大厅”。它不是以计算机技术作为探寻或生成形式的工具来产生形体，而是通过人们的意愿表达来确定建筑的形式。因此该项目不是依赖于任何一种特定的计算技术，而是通过使用多种不同工具而生形。建筑的体量是通过脚本生成的。它能够以一种特定的有旋律的特点使得每一个体块只和其他体块有两个交点。但是这种集合的体块在建模时有一个问题是建筑表面从有规律的平面变换到曲面。

This project for a city of aggregated lobbies is defined not by using computational techniques as form finding tools or form generating tools, but rather through the expression of a will to make form. The project therefore does not rely on any specific computational technique but rather is product of the use of several different tools. The massing of the volumes is generated by scripting, which allows metrically specific features to aggregate in such a manner that each volume only intersects every other volume in two points. The aggregated volumes are modelled as a problem of surfaces that transform from ruled planes to spherical curvatures.

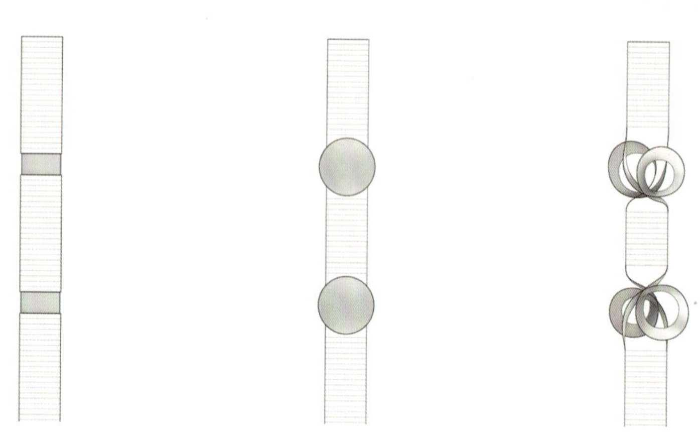

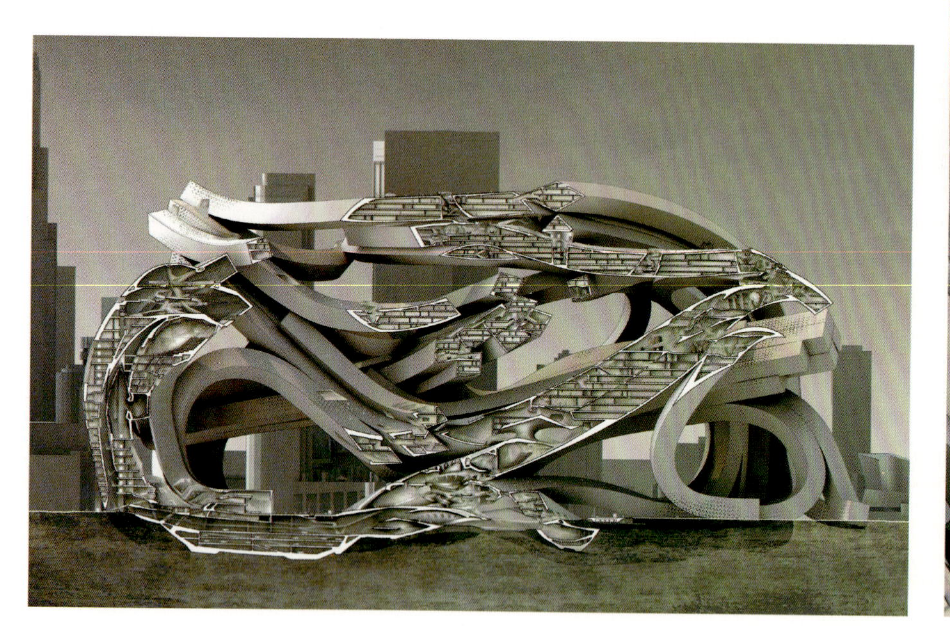

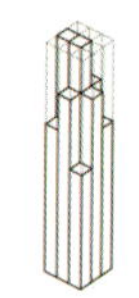

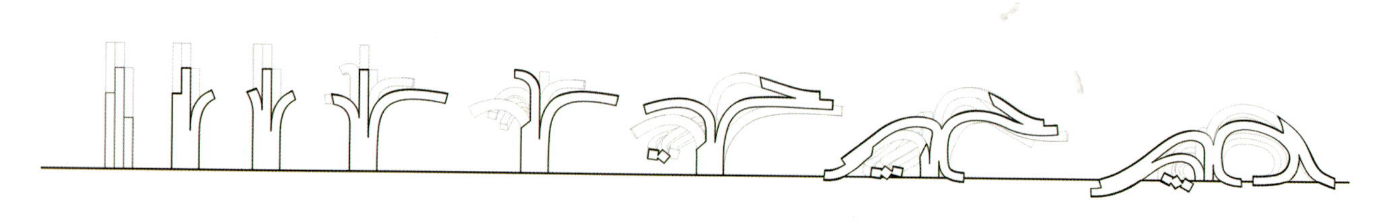

聚合管束 / Aggregated Bundle Tubes

该项目是面向城市的“聚合管束”。这个项目对格莱格·林设计的西尔斯大厦项目进行了一项深入的计算和发展。当数字技术第一次在建筑学科应用时，西尔斯大厦率先应用了这项技术。“聚合管束”是通过使用模拟物理现象的建模技术整合生成的。这些被设计成管状的形体具有一个物理引擎。它们的运转不仅受到几何性约束而且也受到重力的影响。一旦物理引擎产生了管束的聚合，空间品质就以布尔运算的形式产生。

This project for a city of aggregating bundle tubes is a further computational development of Greg Lynn's Stranded Sears Tower project, one of the first projects to have been developed when computational techniques were first developed within the discipline of architecture. The aggregation of the bundle tubes is generated by the incorporation of modelling techniques that are used for simulating physical environments. The bundles are generated as tubes within a physics engine that operates within not only geometrical constraints but gravitational ones as well. Once the physics engine has generated the aggregation of tubes, the spatial qualities are modelled as Boolean operations.

聚合楼层 / Aggregated Floors

该项目是面向城市的“聚合楼层”，它着眼于间隙和虚空间的问题。这个项目使用关联设计技术产生正式的结果。与此同时“楼层”并非基于阵列的影响而是由相关联的隐藏值集合形成。这些楼层是由这些数值共同作用的结果。当变化其中一个隐藏的数值时，楼板就会被连锁反应所影响。这项技术已经被进一步应用在三角形的立面表皮上。每一次楼板改变方向性，三角形的间隙会依据这个变化来做出响应。

This is a project for a city of aggregated floors that looks at formal problem of cracks or voids. The project uses associative design techniques to generate its formal outcome. Meanwhile the floors are not based on array effects but are assembled out of associative hidden volumes. The floors are the intersection of these volumes. When one of the hidden volumes is changed the floor plates are affected by the chain of association. This technique has been further applied on the planar triangulation of the facade surface. Each time the floor plates change directionality, the triangulation of the cracks respond accordingly to this change.

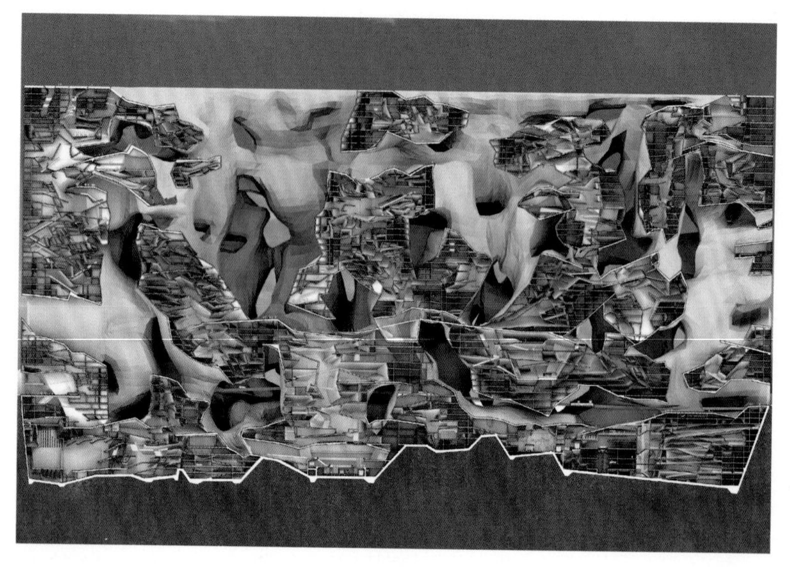

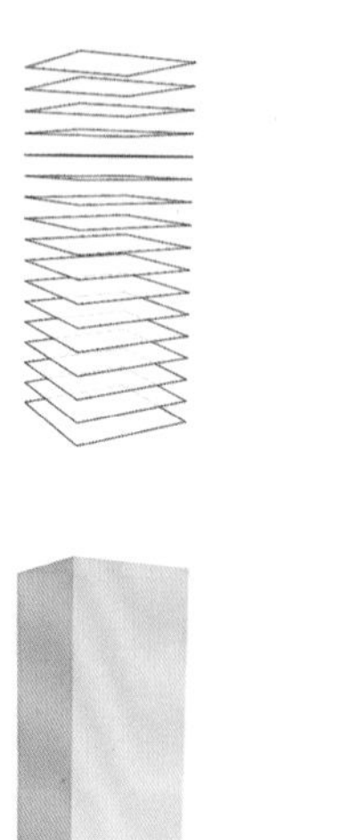

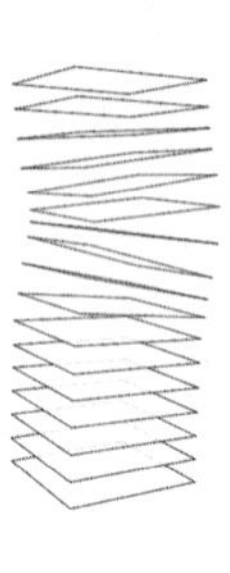

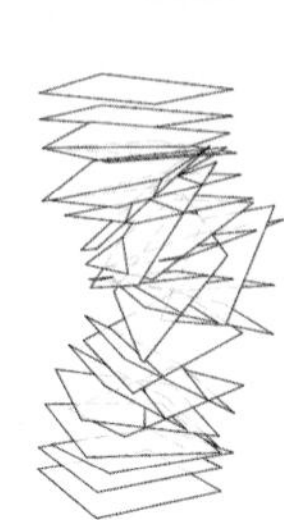

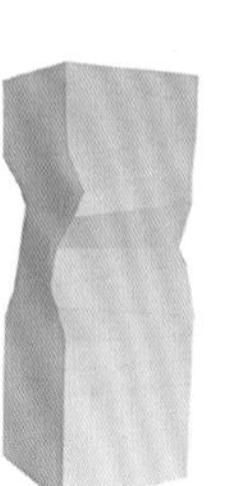

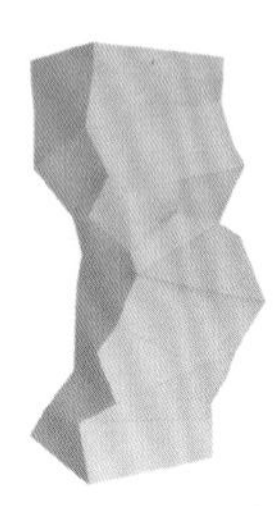

法国巴黎玛莱柯建筑学院 / Paris Malaquais

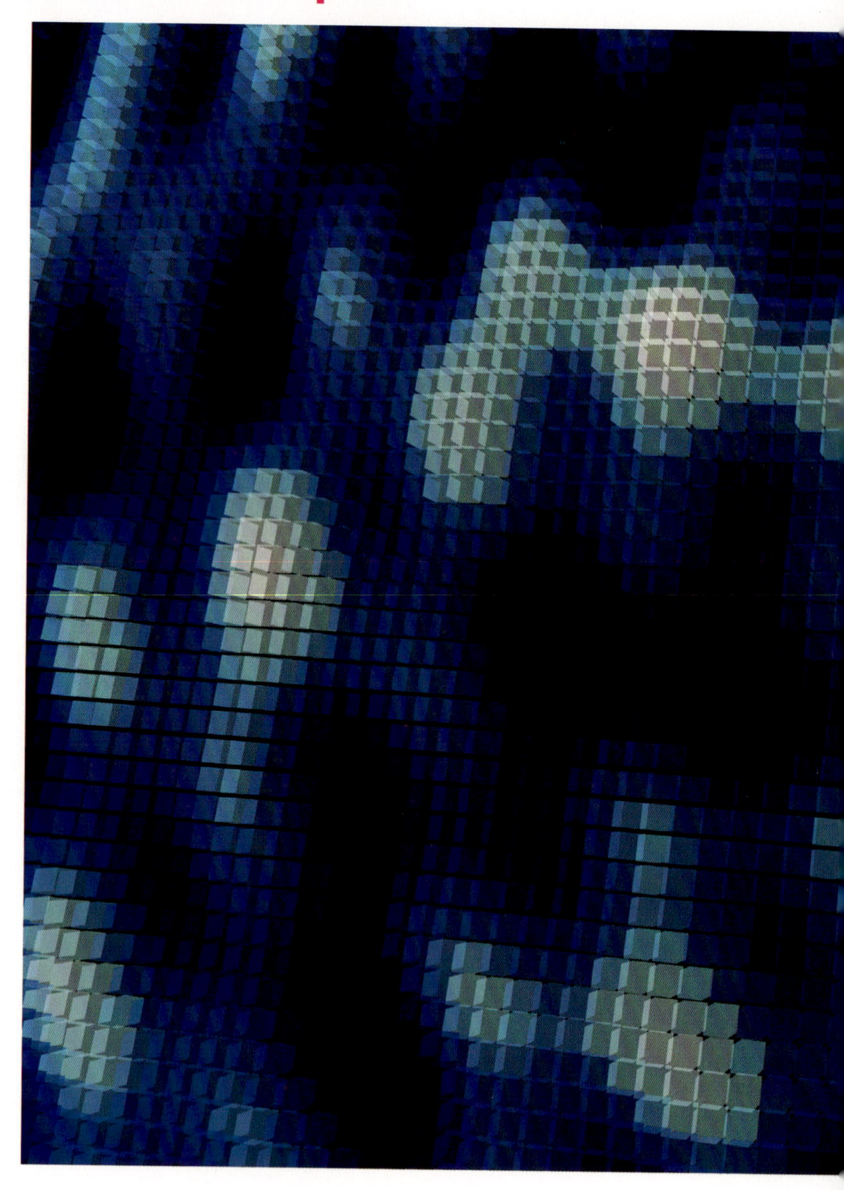

神经设计 / Neuro Design

如果我们以系统化的方式探索空间表现，那么表现的问题需要考虑多个方面，包括感知、判断和操作的阶段。此外，所有的环境数据与人类机体是能够整合的，大脑的神经活动和认知功能也可能被包含在生态信息中。多亏了所谓的“脑机连接”，通过连接大脑活动和建模形式，这项前进中的研究尝试着探索连接与形体生成和演化相关的信息认知的方法以及具有连贯的形态结构组织模式的几何聚集。

If we approach spatial representations in a systematic manner, the question of representation can take on various aspects given the stage of perception, judgment, and operation. In addition to all the environmental data the human body is able to integrate, the neural activity and cognitive functions of the brain may also be included within an ecology of information. By interfacing brain activity and the generative modeling of forms, thanks to the so-called Brain-Computer Interfaces, this ongoing research attempts to explore ways of linking the cognition of information related to the generation and evolution of shapes and geometrical aggregations with coherent models of morpho-structural organizations.

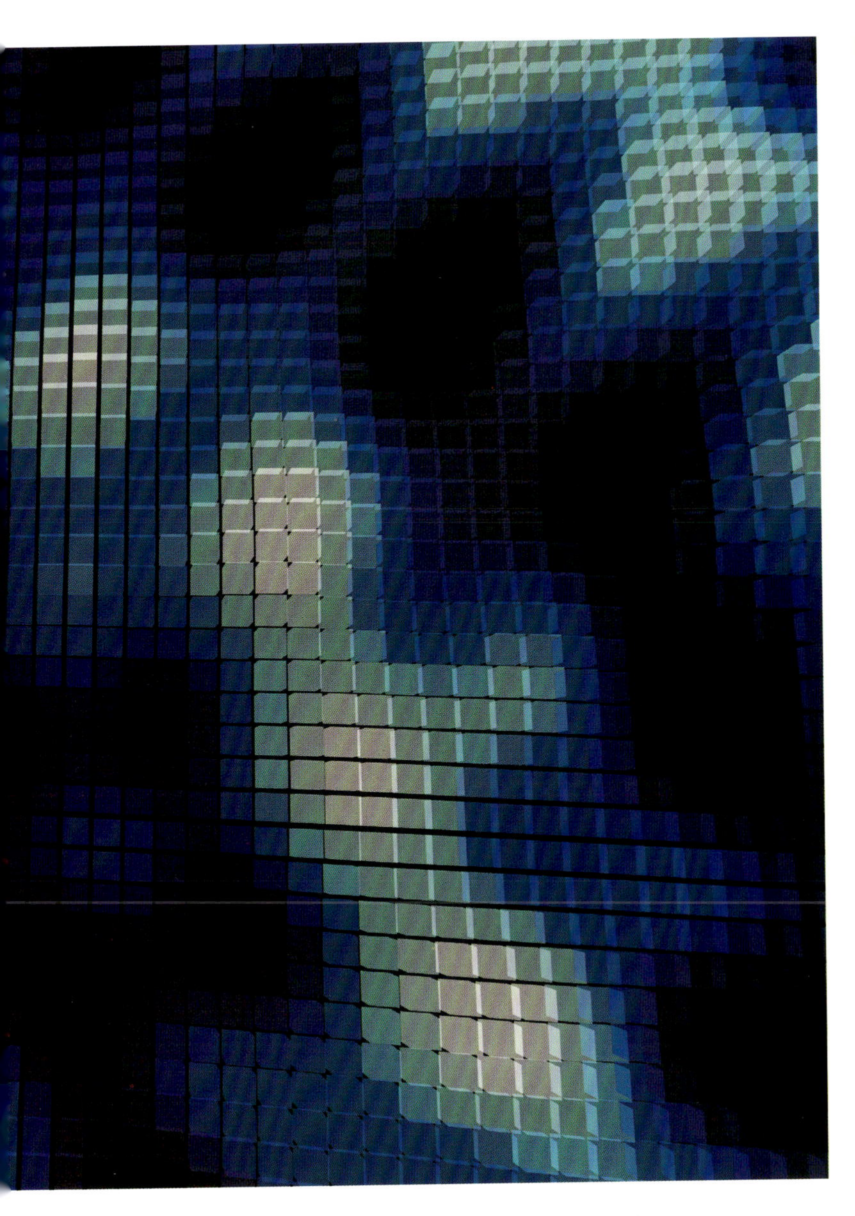

反馈决策 1/
Regressive Resolutions 1

传统意义上，建筑项目的发展被认为在实践、合同和组织方面都是线性和单一方向发展的时尚。在综合项目交付和产品生命周期管理领域，最先进的建筑和工程技术已经逐步具有一种基于传统系统的双重视野。此外，在设计和模型管理方面由数字科学和软件业的飞速发展所带来的其他技术突破，已经使计算机辅助技术在处理这种现象方面产生了方向的改变。例如参数化设计和关联性概念已经解答了在项目发展中不断演变的制约因素的多样性问题和快速适应性的问题。

Traditionally, the development of an architectural project is thought of in a linear and unidirectional fashion practically, contractually and organizationally. State-of-the-art architecture and engineering technologies under the Integrated Project Delivery and Product Lifecycle Management niches have gradually operated a double vision of this traditional system. Furthermore, other technological breakthroughs in design and model management coming from the rapid development of computational sciences and the software industry have actively generated a change of direction in the approach to such phenomena by computer-aided technologies. Parametric design, for instance, and the notion of associativity has answered to variations and rapid adaptation to constantly evolving constraints during project development.

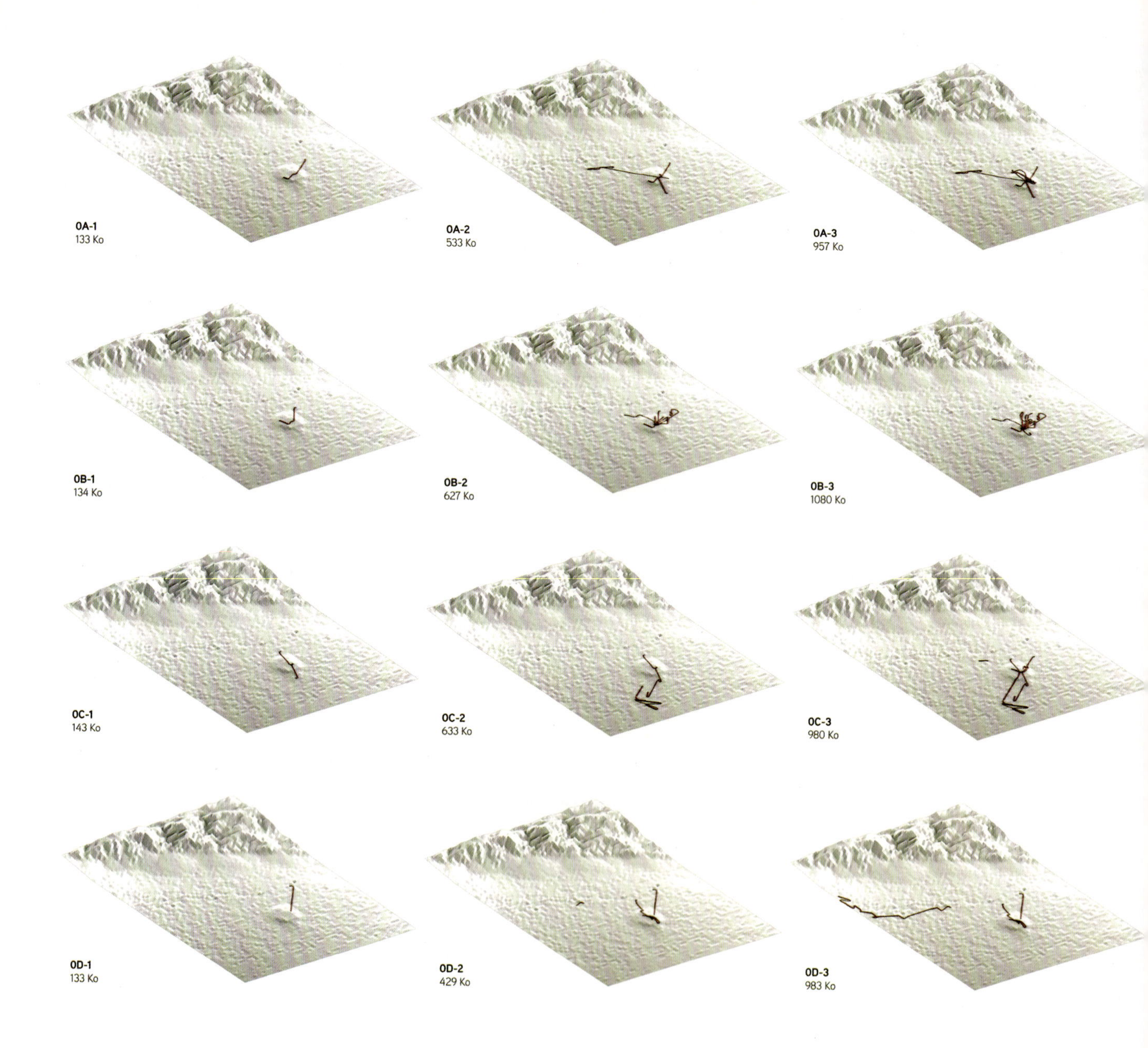
0A-1
133 Ko
0A-2
533 Ko
0A-3
957 Ko
0B-1
134 Ko
0B-2
627 Ko
0B-3
1080 Ko
0C-1
143 Ko
0C-2
633 Ko
0C-3
980 Ko
0D-1
133 Ko
0D-2
429 Ko
0D-3
983 Ko

0A-4
1414 Ko

0B-4
1369 Ko

0C-4
1536 Ko

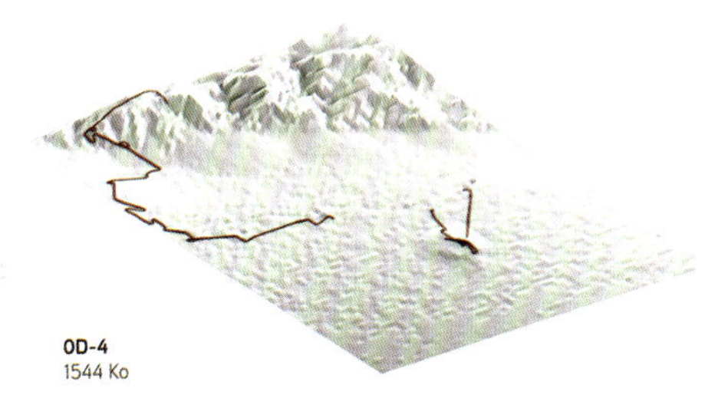

0D-4
1544 Ko

反馈决策 2/
Regressive Resolutions 2

综合性设计和建筑信息建模技术已经通过分享数字模型设计的发展和交付执行减少了处理的数据量。通过把生产模型降低成可交流和快速适应的一系列灵活组件来进行普通的转换，以此在设计思想和建筑实践方面提供了一种回归的解决方法。现在所描述的生产过程可以在两端接近和分化，这种生产通过从任何中间阶段撤回以侧重极端的发展：项目约束表现出的综合性虚拟环境的发展以及其反映的建筑组件的物理聚集。

Integrated Design and Building Information Modeling technologies have reduced the amount of data handled by sharing a digital model of the design through its development and delivery for execution. By reducing the model of production into a communicative and rapidly adaptive set of agile components these general transformations have offered a regressive resolution approach in design thinking and architectural practices. The described production process can now be approached and polarized at both ends, by retracting from any intermediate phase to focus on the development of the extremes: the development of integrative virtual environments of project constraints and its responsive physical aggregation of building components.

德国斯图加特大学 / ICD Stuttgart

复杂张拉曲面的形体学 / Complex Tensile Surface Morphologies

本项目对计算设计研究所的高级人工张拉曲面结构生成与实体化方法研究进行扩展练习。使用 Processing 软件中的点 - 弹簧库来实现基于物理的模拟方法，能够形成复杂的拓扑结构，进而能够被实体化为由钢缆网格与织物组成的混合结构。本项目探讨了生成极其复杂的拓扑结构的可能性，如例子中展示的高级螺旋面，以及在其他课题中研究的多空、多层、有容积的声学几何体。

This work exercises extensive research from the Institute for Computational Design in methods for generating and materializing highly articulated tensile surface structures. Utilizing physics-based simulation methods through the use of a particle-spring library in Processing, complex topologies are formed and subsequently translated for materialization as hybrid structures combining both cable meshes and textiles. The work explores possibilities in generating intensely complex topological arrangements, such as the hyper-toroidal condition shown in one instance, as well as calibrating the porous, layered and volumetric geometries in a multi-cell arrangement for acoustic modulation, in the other set of studies.

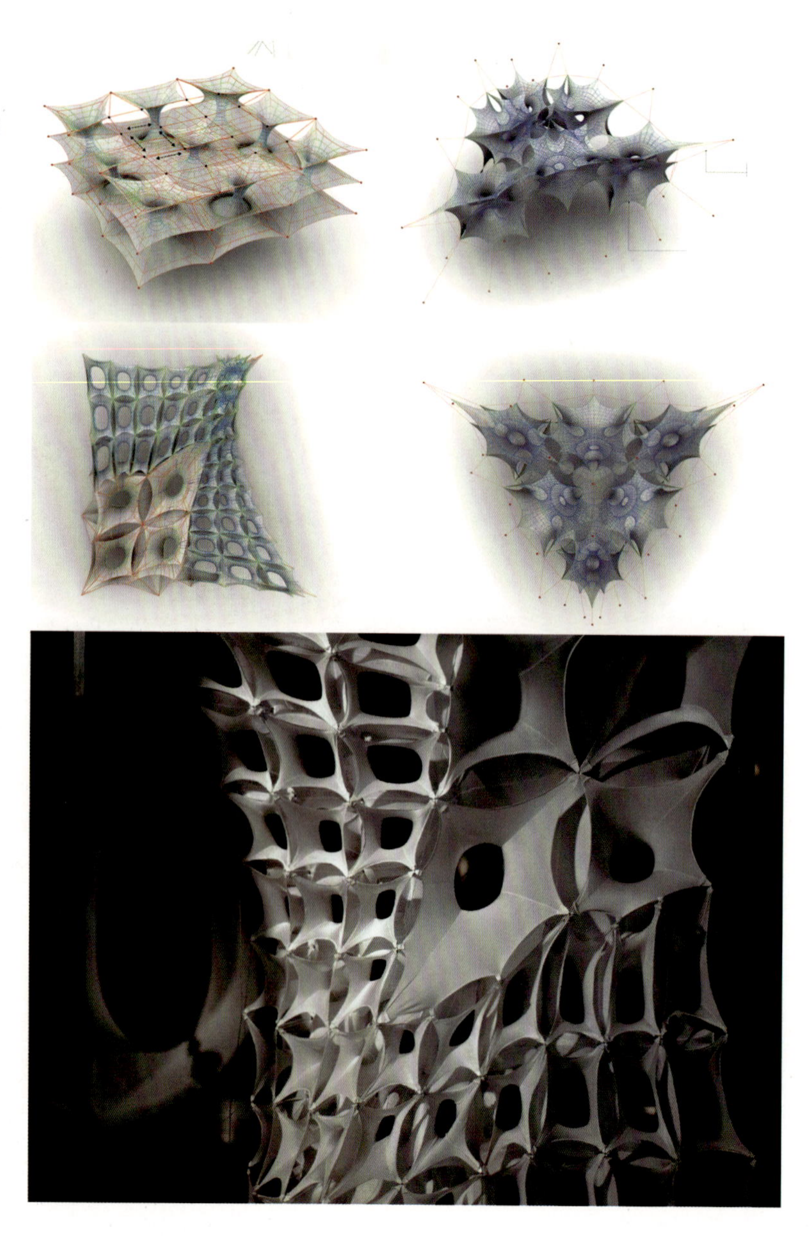

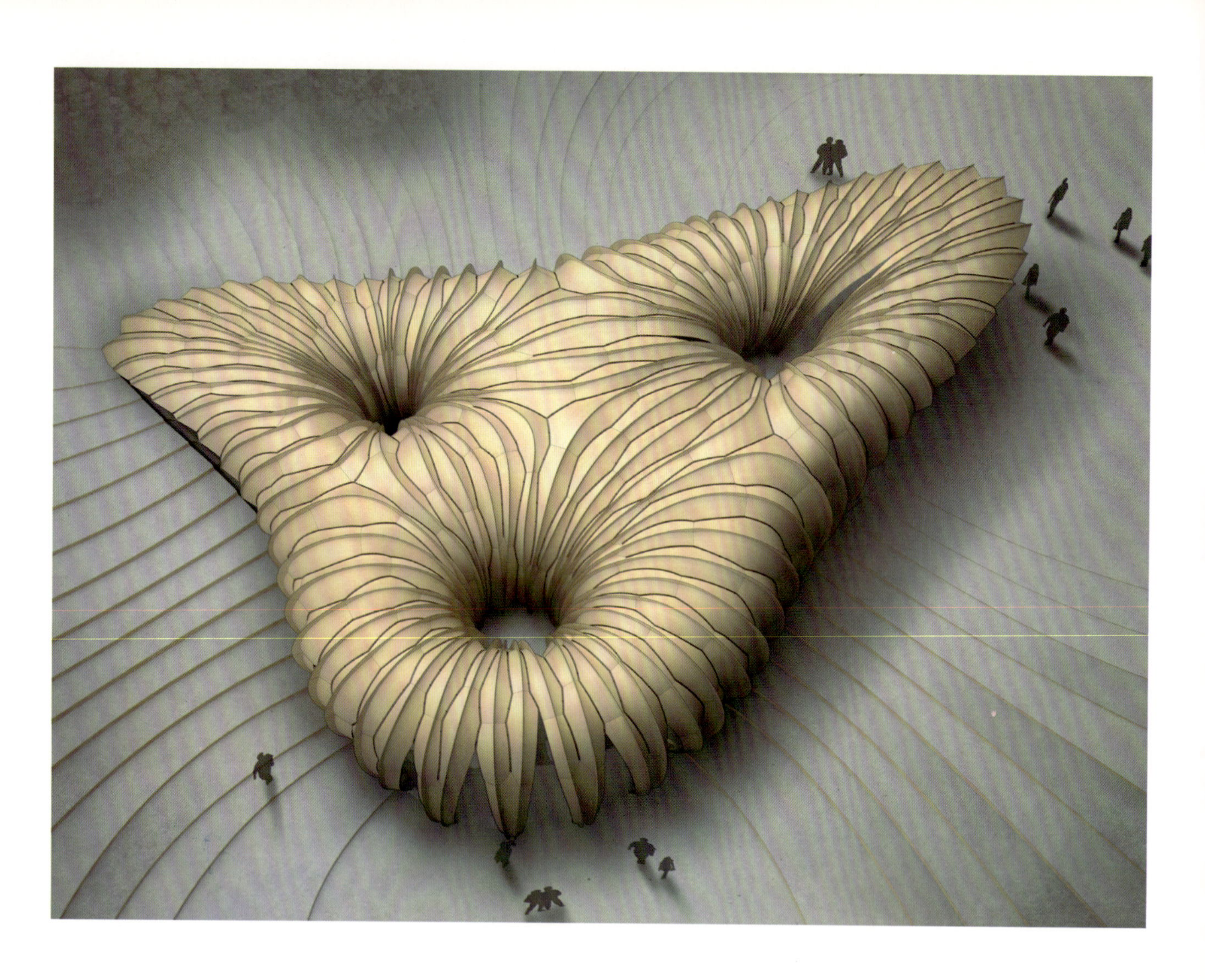

连接智能 / Connecting Intelligence

“连接智能”项目的目标是探讨机器人建造在木结构中的建筑潜力，这种建造基于可变形的弯折纤维片，使用机器人手指节点进行建造。当前计算设计和数字建造的发展提供了与经典自上而下的建筑设计步骤所相反的交互式设计方法。本项目基于机器人的高运动学自由度，这能够为复杂并极富表现力的木板结构单一材料节点的实现提供可能性。在计算设计过程的内容中，这些节点在弯曲可变结构体系的开发中被采用。最终这套系统的建筑应用和相关的结构性能被强调。

Connecting Intelligence aims to investigate architectural potentials of robotic fabrication in wood construction based on elastically bent timber sheets with robotically fabricated finger joints. Current developments in computational design and digital fabrication propose an integrative design approach contrary to classical, hierarchical architectural design processes. The project is based on the robot's high degree of kinematic freedom, which opens up the possibility of complex and highly performative mono-material connections for wood plate structures. In the context of a computational design process, these connections are employed in the development of bending-active construction system. Finally the system' s architectural application and related structural performance is addressed.

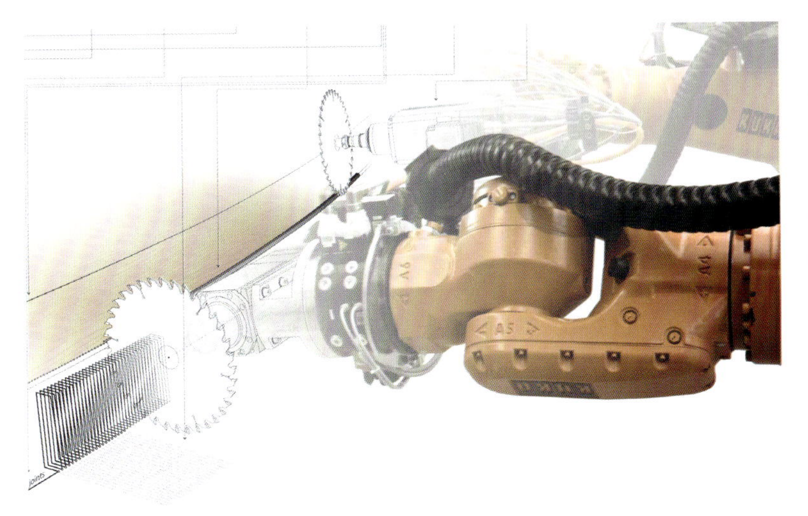

木质复合形态 / Wood Laminate Morphologies

本项目是基于对复杂织物结构和网格形态（如预应力张拉结构）的方法与实体化研究进行的。本项目名为差异化的木质薄片形态学，目标是对轻型张拉钢缆网络结构进行操作，将线性网络元素物化为一系列木质胶合薄片曲面元素。为了实现这一目标，木质元素依据它们的几何特性和层次特性被提取出参数，从而有了最优化的可变性。经过经验测试一系列拼合形态被确定下来，它们能够为可变张拉结构产出合适的可变性与刚性。

This project elaborates upon significant research into methods and materialization of complex textile and mesh morphologies as pre-stressed tensile structures. The project, entitled Differentiated Tensile Wood Laminate Morphologies, seeks to operate as a lightweight tension-active cable-mesh structure, opting to materialize the matrix of linear mesh elements into an array of laminated wood-veneer surface elements. To do so, the wood elements are parameterized in their geometry and degrees of lamination to allow for optimal flexibility. Through empirical testing, a range of tile morphologies is established, that are capable of producing the appropriate relationship of flexibility to stiffness in registering a tension-active structure.

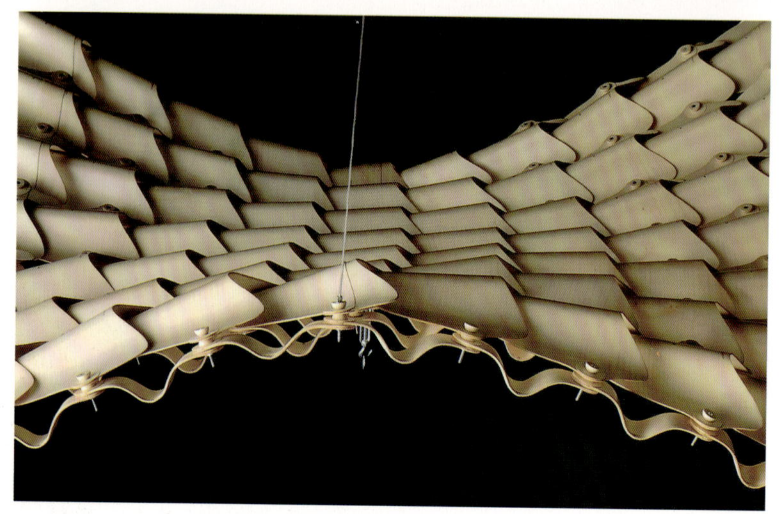

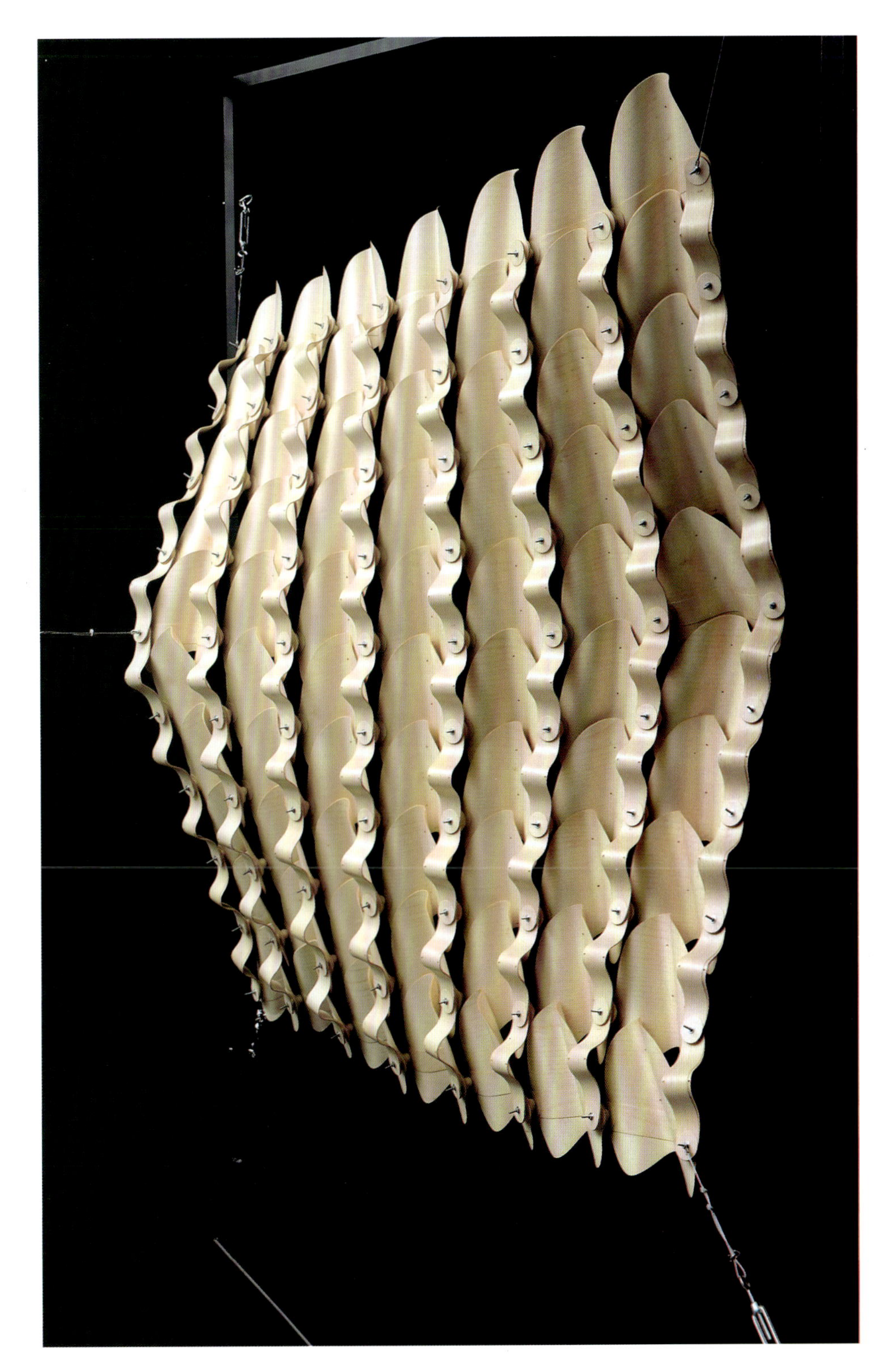

瑞士苏黎世联邦理工大学建筑学院 / ETH Zurich

砌砖逻辑 / Brickolage

这个项目的主旨是设计一个有关构造的生成性计算机模型，并让库卡机器人使用“加气混凝土”材料来建造完成。学生致力于如何使用计算机去生成一个具有审美的结构，而不是手工塑造。设计思路是用 Processing 软件编写一个“反应 扩散”的算法，生成一个三维的无需支撑的结构。这一结构的“几何形”会再用工具进行加工并依据材料的状况，做进一步的处理。整个结构包括大约一千块砖，使用机器臂及其附属的线锯来完成组装。

The brief for this project was to design a temporary structure using generative computer modeling and fabricate it out of aerated concrete using a Kuka robot. The students were encouraged to think how they could use the computer to generate an aesthetically appealing structure, which could not be done manually. The idea was to use a Reaction - Diffusion algorithm, programmed in Processing, to generate a 3 - dimensional free - standing structure, whose geometry was further processed based on tooling and material constraints. The whole structure was composed of around 1000 bricks, was assembled with the aid of a robotic arm and an attached jigsaw.

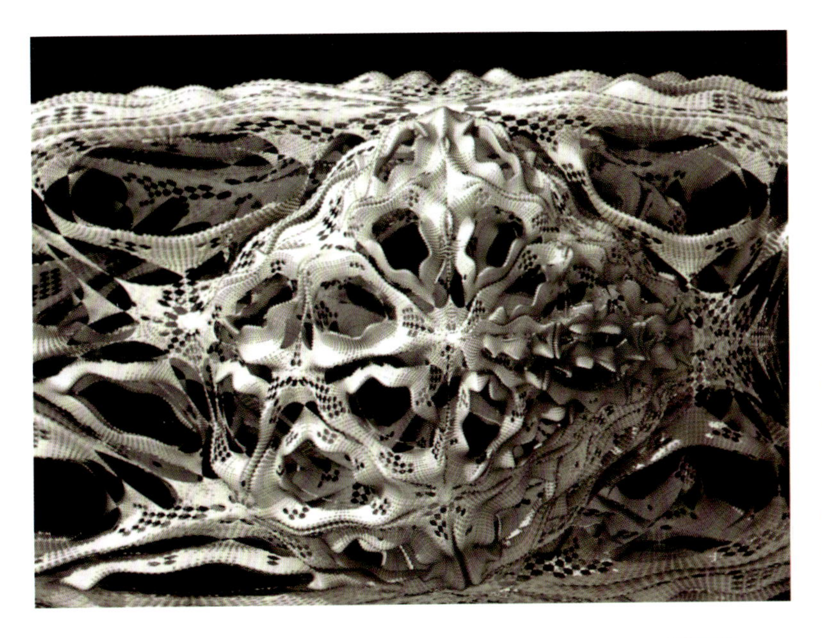

高度解析 / High-Resolution

这个项目通过计算机的途径来设计神圣的空间。通过分析教堂、清真寺和庙宇的几何式样，把它们的法则进行推敲并将其写入形式的算法中。项目运用了“造型文法”的方法论，这是一种生成设计的正规语言，它的开发可追溯到早期的计算机辅助设计时代。今日的电脑已具有巨大的运算能力，从而产生了一种全新的由计算机生成的建筑类型。所有的结构体系都可以使用电脑进行计算获得。而获得的这些极具复杂性的空间是使用传统工具所难以达到的，甚至是人们过去所根本无法想象的。

This project used a computer - based approach to design sacred spaces, based on an analysis of the geometric patterns of cathedrals, mosques and temples, whose rules were analysed and then encoded in the form of algorithms. The project worked with the methodology of 'shape grammars', a formal language for generating design that was initially conceived in the early days of CAD. With today's computers - which are millions of times more powerful - a new level of computer - generated architecture can be achieved. Entire families of structures can be calculated by the computer, whose spatial complexity would be unachievable - and perhaps even inconceivable - using traditional means.

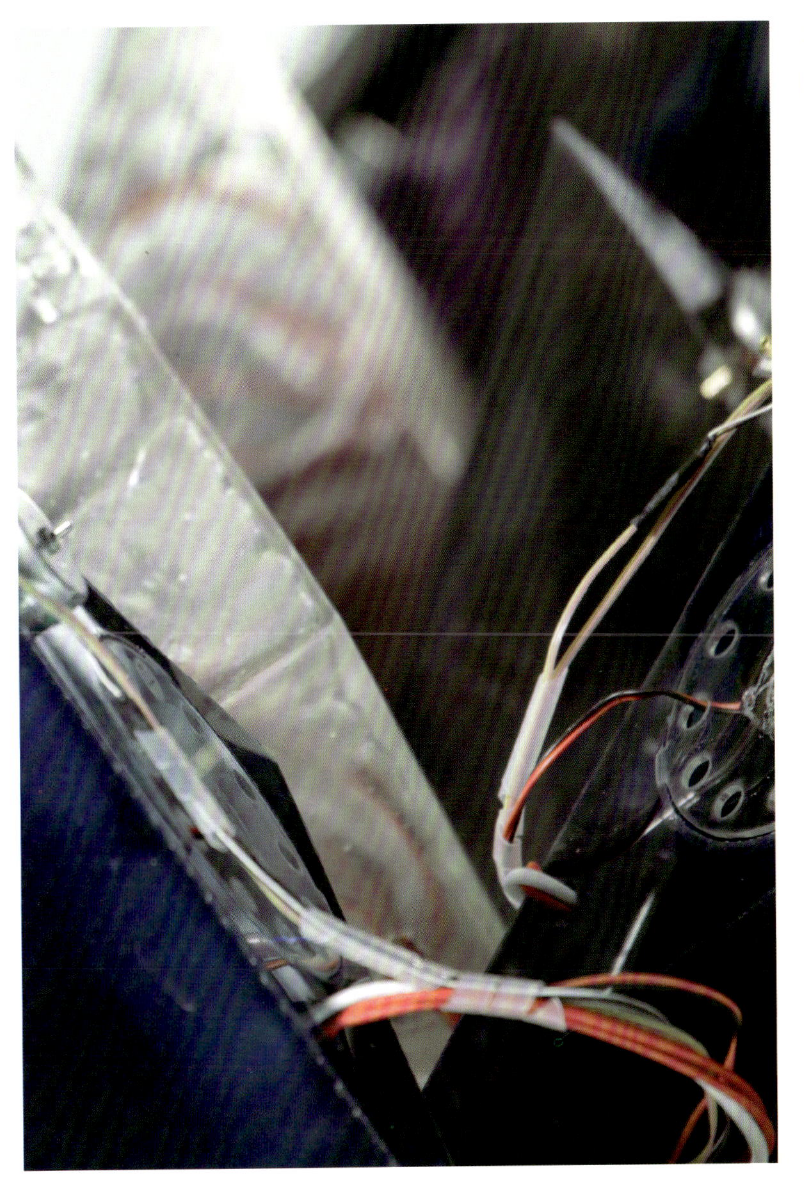

共振 / Resinance

共振指受到简单的生命有机体的行为的强烈影响，尤其是在形成细胞集落方面。这个设计由具有生态学功能的单元体组成。这些单元体既可以自主地运行，也可以与其相邻的单元体一起协同合作。这个设计包含了 40 个可以抖动和振动的“活性元素”。当人们碰触它，它会逐渐的改变自己的颜色来作为响应。这些单元体通过精心设计以组成特别的群簇，产生出不同行为。并且单元体会把自己当前的状态传递给每一个自己周围的单元体。这样输入的触觉不但改变了被触碰的单元，也会在安装的整个网络上进行传递，形成类似于集群的行为。

Resinance was influenced strongly by the behavior of simple organic life forms, in particular the formation of cellular colonies. It resembled an ecology of functional units that could both work autonomously but also in coordination with their neighboring units. It consisted of 40 active elements that shivered and vibrated, and gradually changed their color in response to human touch. These units both choreographed the behavior of the particular cluster and transmitted the current state of each of its neighbors. Therefore the tactile input not only changed the touched element but was also transmitted throughout the whole installation in a networked, swarm like behavior.

澳大利亚皇家墨尔本理工大学 / RMIT

FabPod 项目 / FabPod

FabPod 原型是声学的八音符。它是研究建筑几何与声音关系的成果，将声学设计、建筑、数字建造的研究者聚集在一起。FabPod 展示了一个可以快速、半自动化生成形态、几何连接和材料分配的新奇的设计系统和数字流程。声学模拟作为这个流程的一部分，提供了对变化的设计参数的快速反馈。这种方式使声学性能可以成为决定性的设计因素，并且设计师可以精确地理解究竟他们的决定如何影响这个性能。

The FabPod prototype is an acoustically tuned meeting space for eight. It is an outcome of research into the relationship of architectural geometry and acoustics, bringing together researchers in acoustic design, architecture, digital fabrication and manufacturing. The FabPod showcases a novel design system and digital workflow allowing for the rapid, semi automated generation of form, geometric articulation and material distribution. Acoustic simulation as part of this workflow provides rapid feedback in response to changing design parameters. In this way the acoustic performance could drive key design decisions and designers developed a sophisticated understanding of how their decisions influenced the performance.

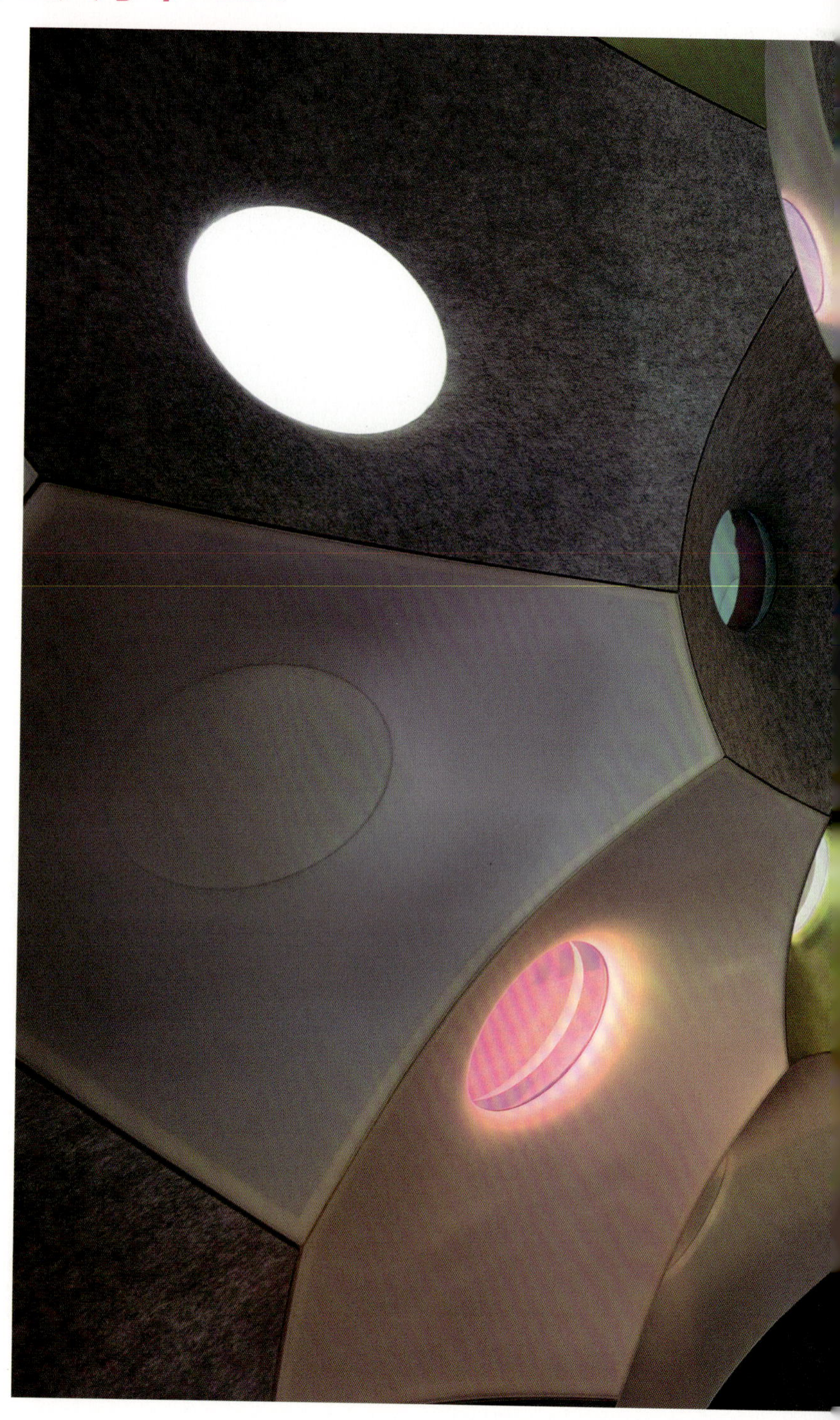

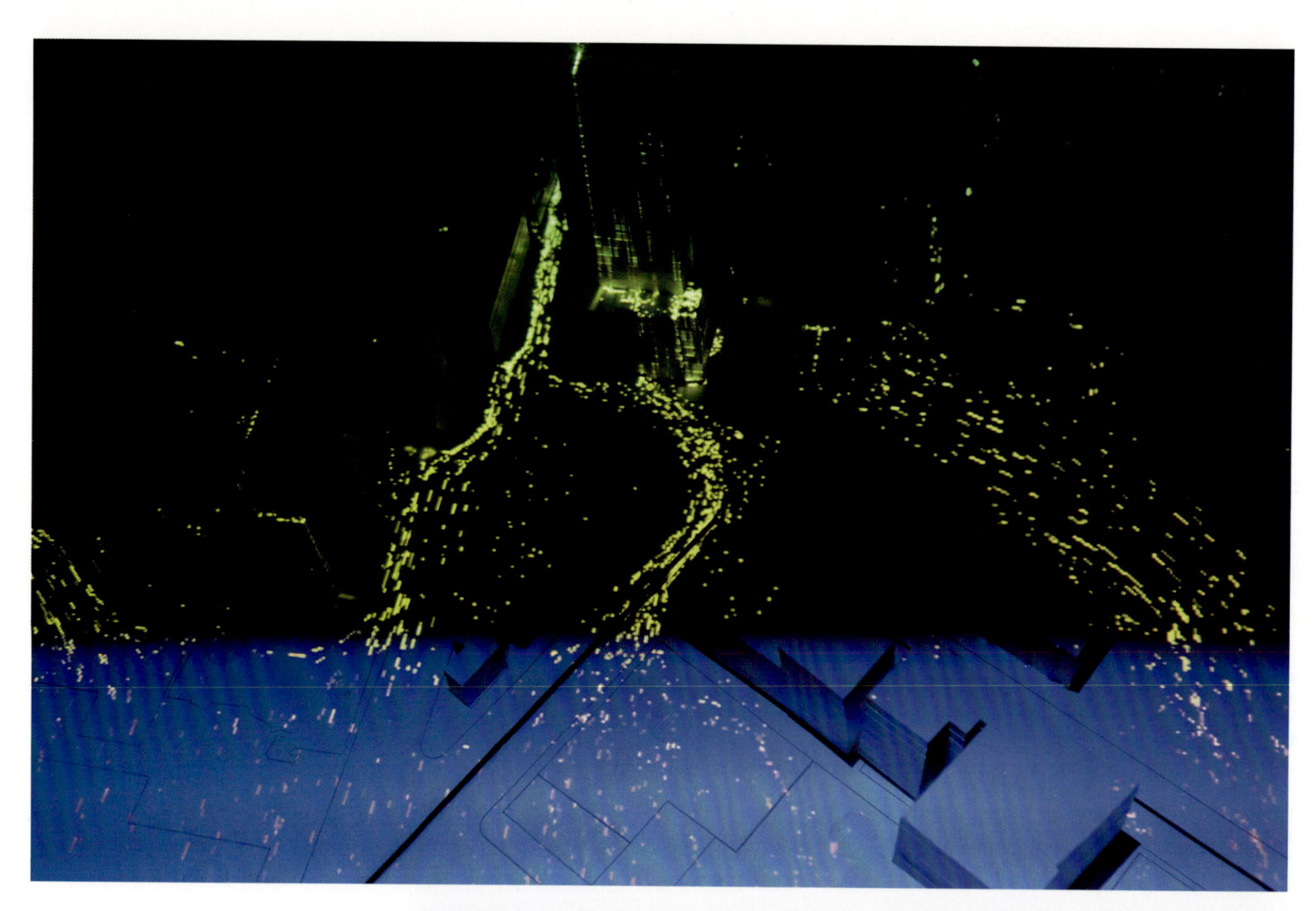

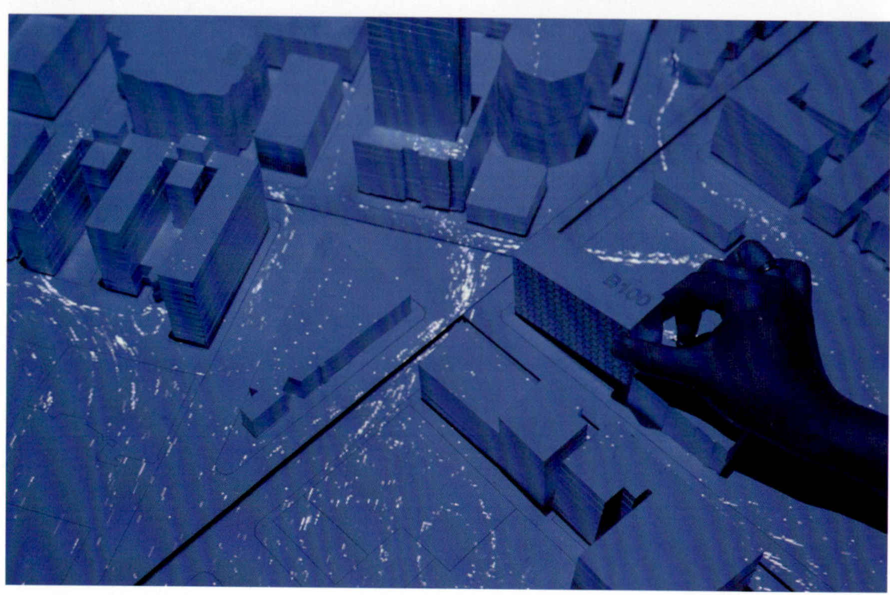

有形的协同工作台 / Tangible Teamwork Table

这个项目提供了一个新的设计流程方式，使得实体模型和虚拟模型通过许多分析数据集和实时数据流实时关联和通知。设计师应用不同的建筑街区（由泡沫、纸或卡片制成）研究任意给定城市的日照和风环境影响。通过严谨的对一系列街区的操作，使用者可以发现立体的城市配置有可能减缓或消除区域里的风，并且满足日照的需求。

This project offers a new approach to design workflow, where physical model and virtual model can be interconnected together and informed in real-time by multiple analytical datasets and live data streams. Designers could use various building blocks (made of materials such as foam, paper, or cardboard) to explore the effect of any given building design on the site in regard to local sun and wind conditions. Through careful manipulations performed on the set of blocks, users could find urban volumetric configurations that could potentially slow down or cancel the wind in the area and match the desired solar radiation.

复合代理 / Composite Agency

这个复合的原型是将结构、表面、装饰压缩成单独的相互依靠的集合建筑设计项目。该项目通过微观尺度计算设计代理的互动引起宏观尺度复杂事件的行为学方法来进行设计。橘色的代理主体自组织地创造了在复杂纤维表面内部的结构镶嵌体。这些主体部分在多余的装饰与结构需求之间先化并生成形体。复杂的表面和代理主体在皇家墨尔本理工大学由机器进行加工。

This composite prototype is an architectural studio design project that compresses structure, surface and ornament into a single mutually dependent assemblage. The project is designed through a behavioral methodology where the interaction of computational design agents at the micro-scale can give rise to the emergence of a more complex order at the macro-scale. The orange agent-bodies self-organise to create a continuous structural inlay within the composite fiber surface. These bodies negotiate between ornamental excess and structural necessity in articulating the form. The composite surface and agent - body formwork were robotically fabricated at RMIT.

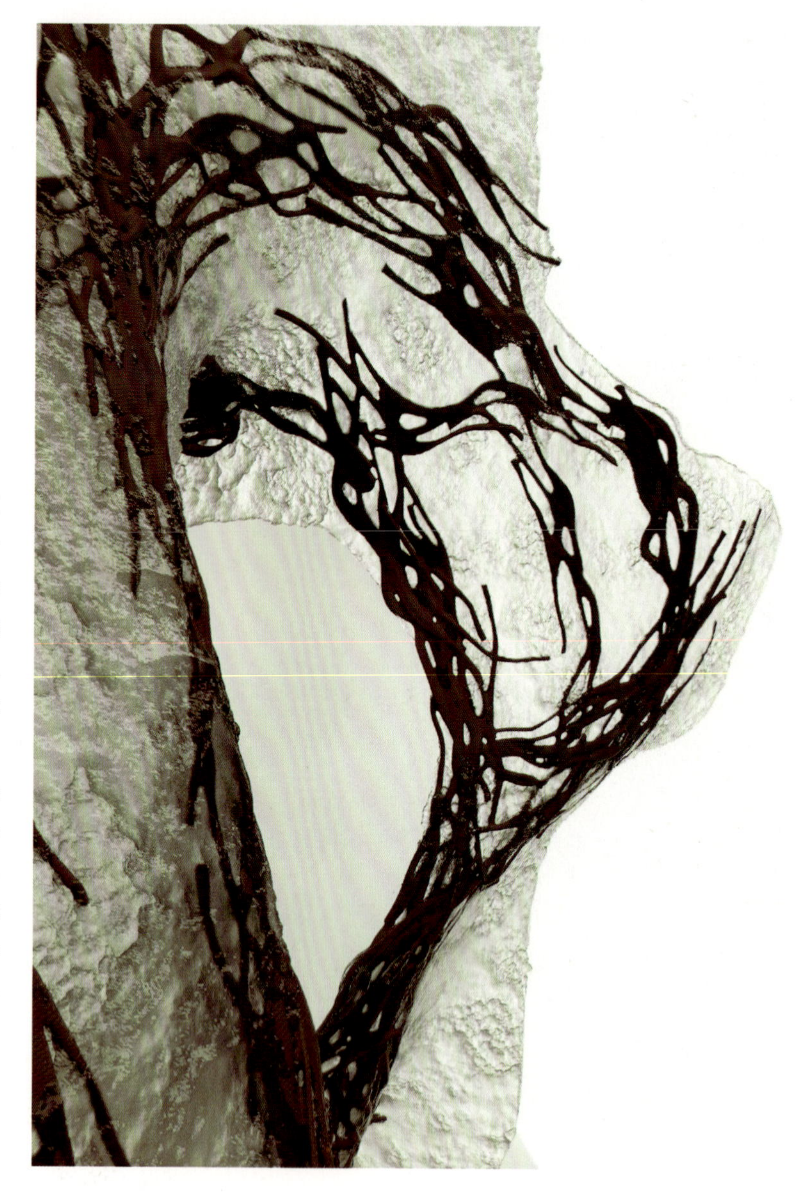

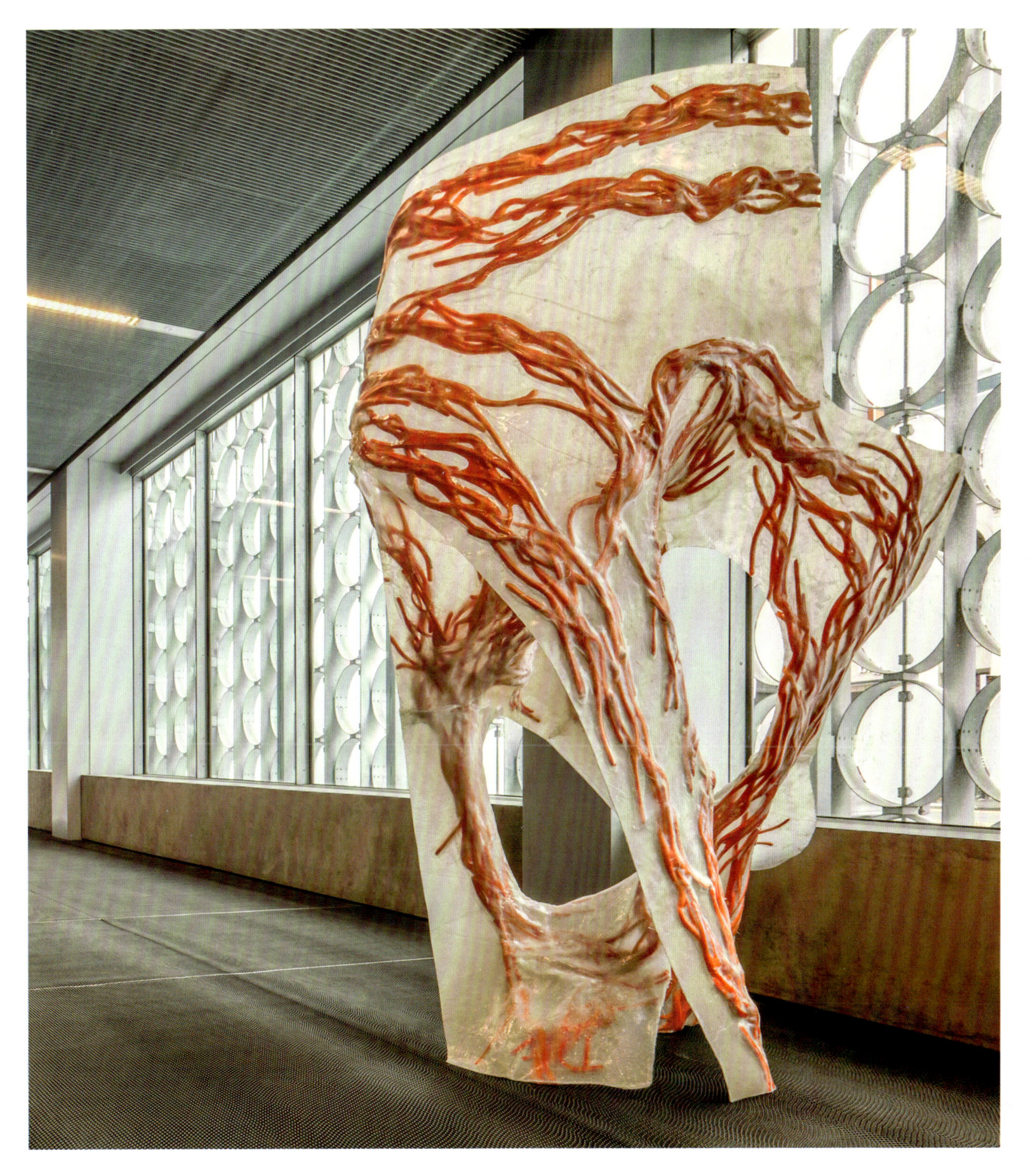

中国香港大学建筑学院 / HKU

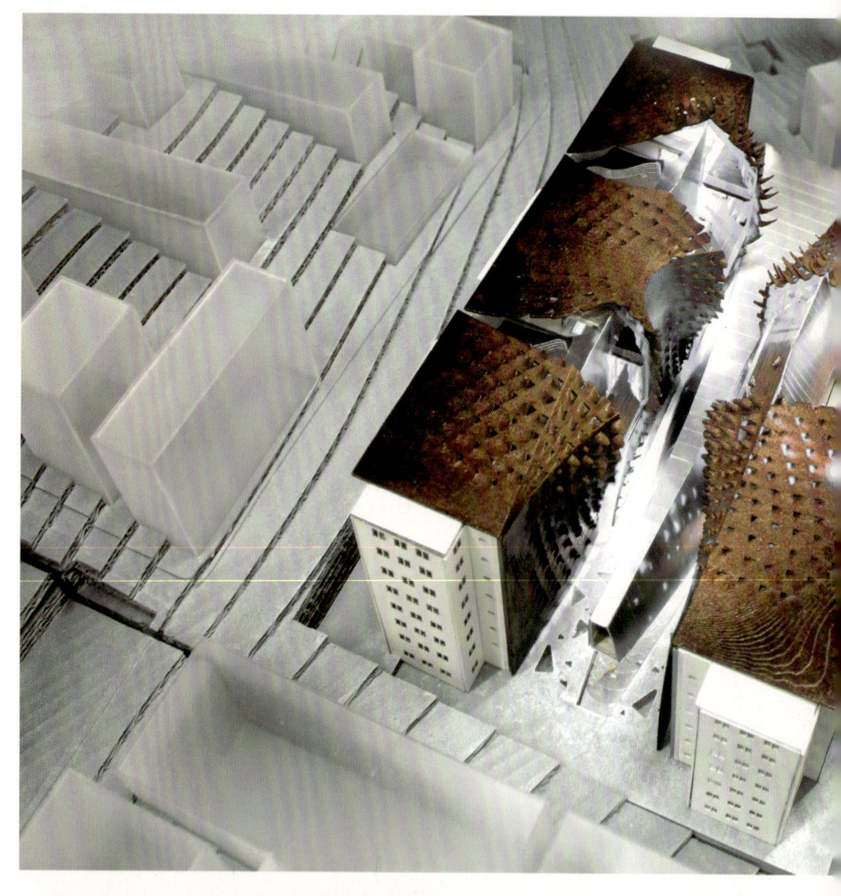

包容的闭塞性 / Inclusive Occlusion

通过一个自适应更新项目的视角，该方案使用一种阻断的策略来探索关于密度和聚合的概念。基地位于香港前已婚警察宿舍区，四个新的体量被插入到现存建筑中去，提升了建筑的密度和潜在的分离感。这四个体量、现存环境和光圈纹理状的新覆盖层（这协助保护了原有的建筑）组成横剖面几何体，其被数字建模及广泛原型化，这用来理解“度”的概念，连同这些“度”他们能制造一个有凝聚力的整体。远处观看几乎难以察觉，该方案同时具有闭塞性和包容性。

Through a strategy of occluding this project explores ideas of density and cohesion through the lens of an adaptive re-use project. Sited in the Former Police Married Quarters in Hong Kong, four new volumes are inserted within and between the existing buildings, increasing the building's density and potential disjunction. The cross sectional geometry of the four masses, original buildings and aperture - texture of the new cladding (which assists in conserving the original buildings) were digitally modeled and prototyped extensively to understand the degree to which together they produce a cohesive whole. Barely perceptible from afar, the project is both occlusive and inclusive.

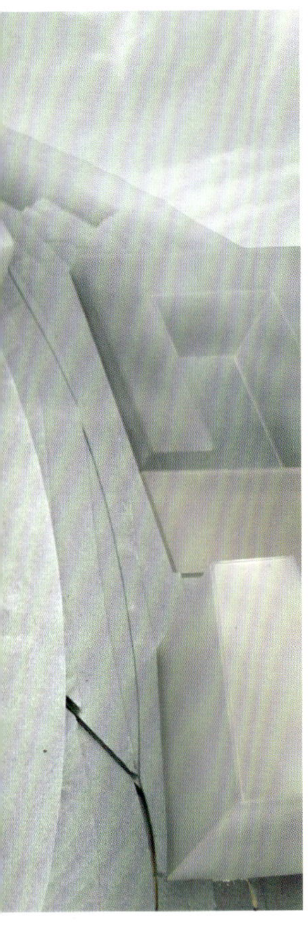

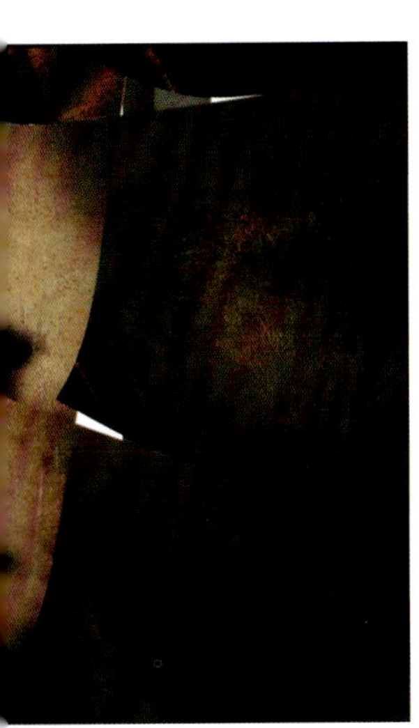

触发器（拱门）/ Trigger (Arch)

这个基于现实的游戏关注于视频游戏软件 UDK 的使用（虚拟开发工具箱）。在设计一个塔时使用了“触发器”——手动控制的代名词，用来转换“或 / 且”动作的传感器，其在建成环境中能够动态性地控制建筑。建筑不同楼层拥有不同的主题——例如旋转的齿轮和倾斜的地板——玩家通过一个障碍后会遇到一个经典的视频游戏，利用触发器来根据他们的需要改变环境。由此产生的项目结果如同充满娱乐性、互动性且身临其境的视频游戏，其允许参与者去体验改变，在这个高度机械化的塔里就如同身处数字城市中。

This reality game focuses on the use of video gaming software UDK (Unreal Development Kit). A tower is designed using "triggers" - manual controls synonymous to switches and/or motion sensors in the built world which control the dynamics of the building. With a different theme at each level of the building - such as rotating gears and tilting floors - players pass through obstacles one would encounter in a classic video game, utilizing triggers to change the environment to their needs. The resultant project is a fully playable, interactive and immersive video game that allows participants to experience the changes in a highly mechanical tower situated in a digital city.

季节表皮 /
Climate Skin

为了应对复杂的气候和环境力量对建筑维护施加的作用，该方案旨在应对各种偶然性并嵌入一套智能性系统。运用先进的参数化设计和模拟工具，设计研究工作从媒介转移到自动计算控制系统，其使用传感器和执行器去处理和计算那些复杂的输入参数，例如气候信息、能源消耗、美学及公共的价值、室内舒适的环境现状及其他关注点。最终，建筑表皮的原型整合了这些诸多方面，如太阳辐射及温度、通风、污染等，创造了一种智能化、反馈化的表皮，并具有可识别性，与地域文化特征相结合。

Responding to the complex climatic and environmental forces acting upon an architecture envelope, this project aims for diverse contingencies to be managed and embedded in a set of performative systems. Using advanced parametric design and simulation tools, the design research work migrated across media to automated computational control systems using sensors and actuators to process and act upon complex inputs of climatic information, energy consumption, esthetics and communal value, environmental conditions of interior comfort, among other concerns. Finally, the building envelope prototype integrates these aspects of solar radiation and temperature, wind and ventilation, pollution, etc, to create an intelligent, responsive façade, with identifiable local culture characteristics.

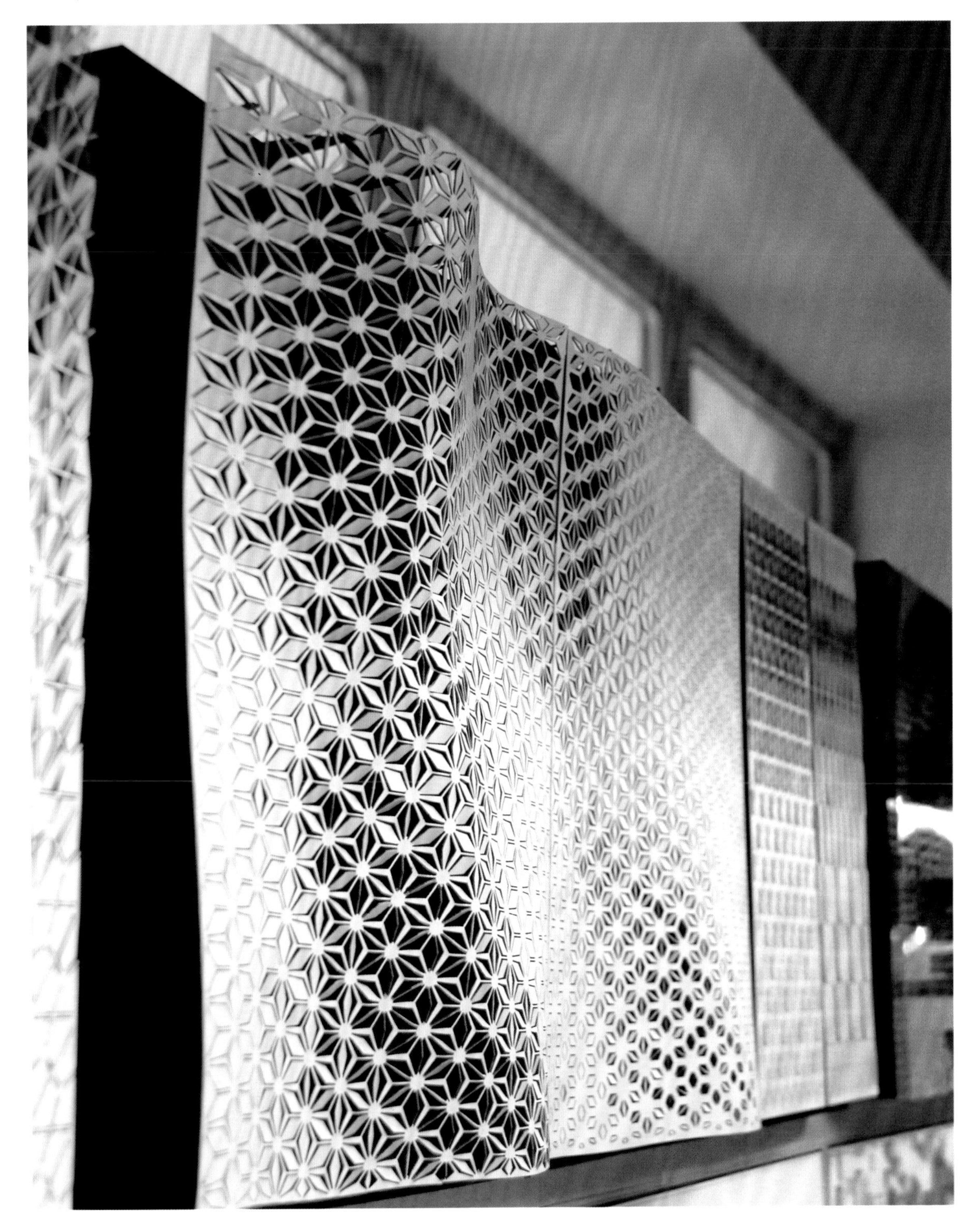

中国华中科技大学建筑学院 / HUST

青年旅社设计 / Youth Hostel Design

探索新的结构是这次作业的核心，从植物中寻得了新结构的启发。植物的结构系统包涵了承重、输水、养分供给等各个方面的综合系统。而人类所创造的结构系统却仅仅用来承重。创造一个不同于传统结构的新系统，不仅解决结构问题，也带给人全新的空间体验。在植物里，各个层级没有明确的划分，每个部分都与上下层级有着紧密联系。在新的结构中，无论承重钢脊的分支，还是与膜结构的连接，结构成为一个整体。

Plants provided the inspiration behind this project, which sought a new system of structure as its core. Plants have a complex system including their physical structure, and their system of circulating water. However, man-built structure typically concerns with load-bearing and lateral force only. Thus the project is aiming at creating a new type of structure which is different from typical one: While providing support, the structure is also developed to create a magnificent spatial experience. There is no clear distinction between different parts of plants, and no clear classification among their veins. They are all just sections of a complex system.

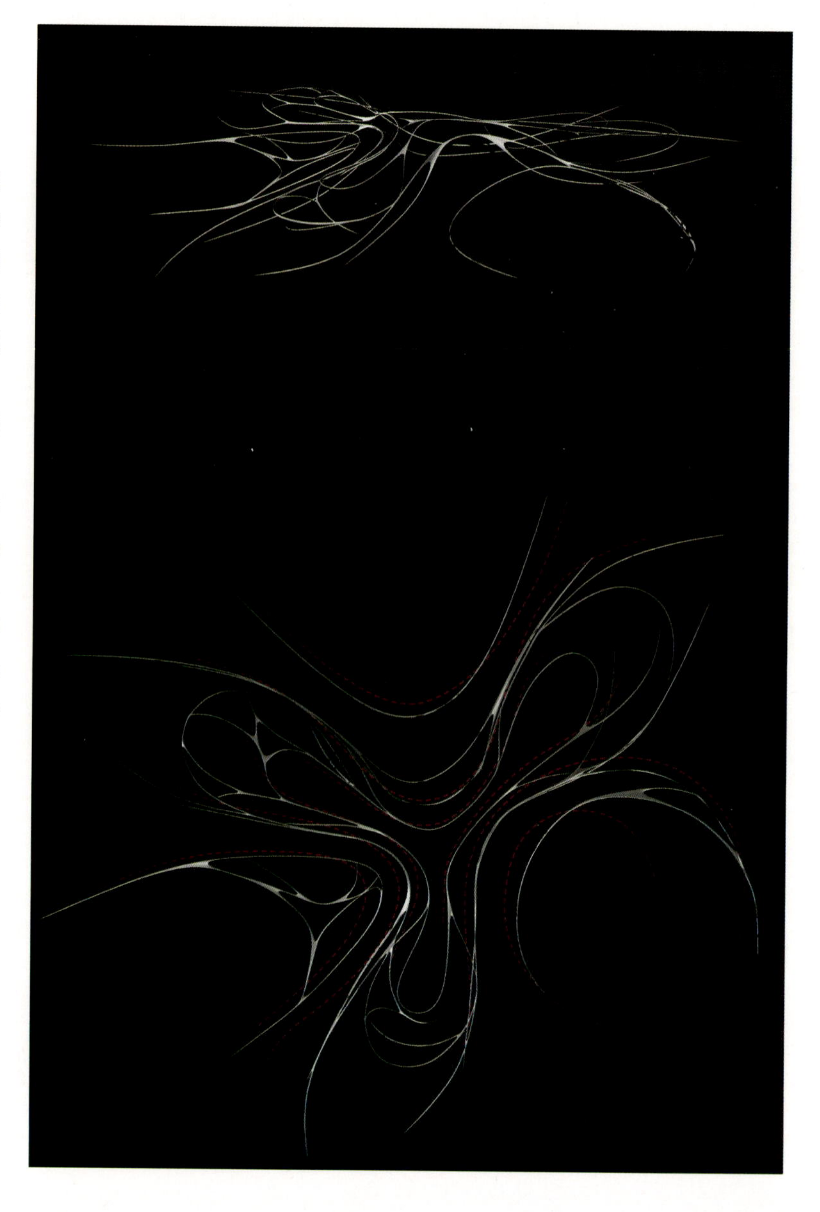

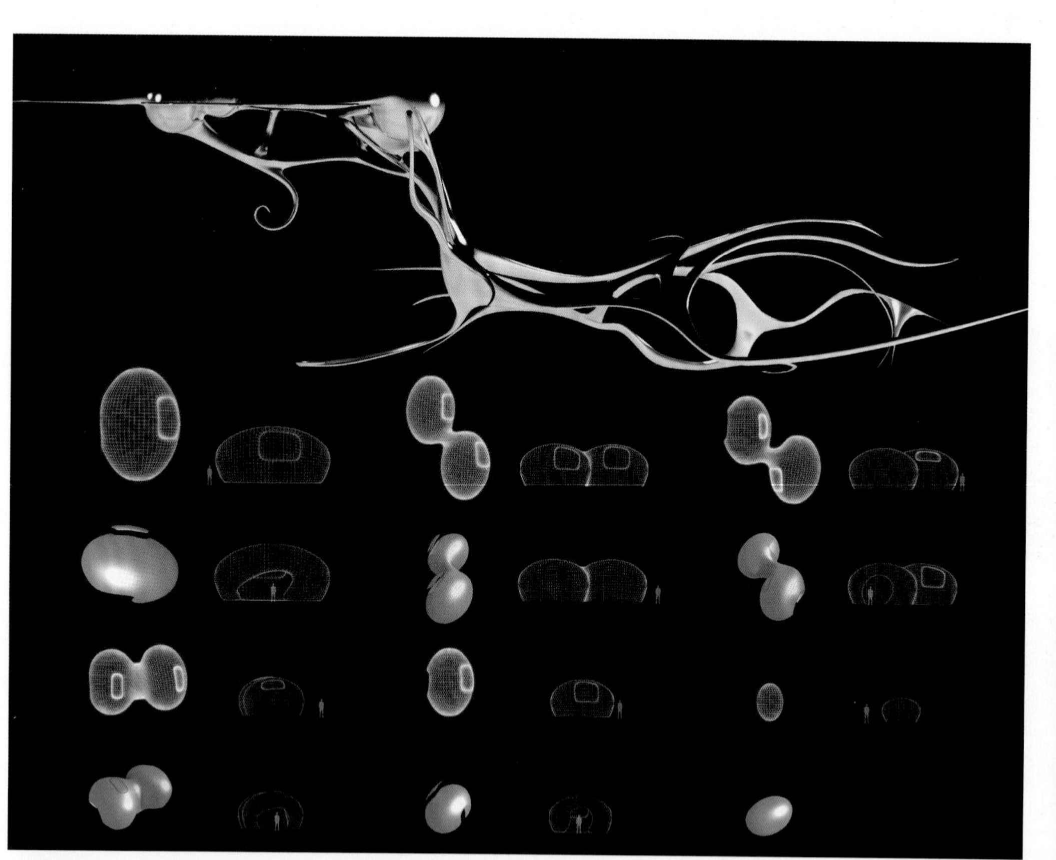

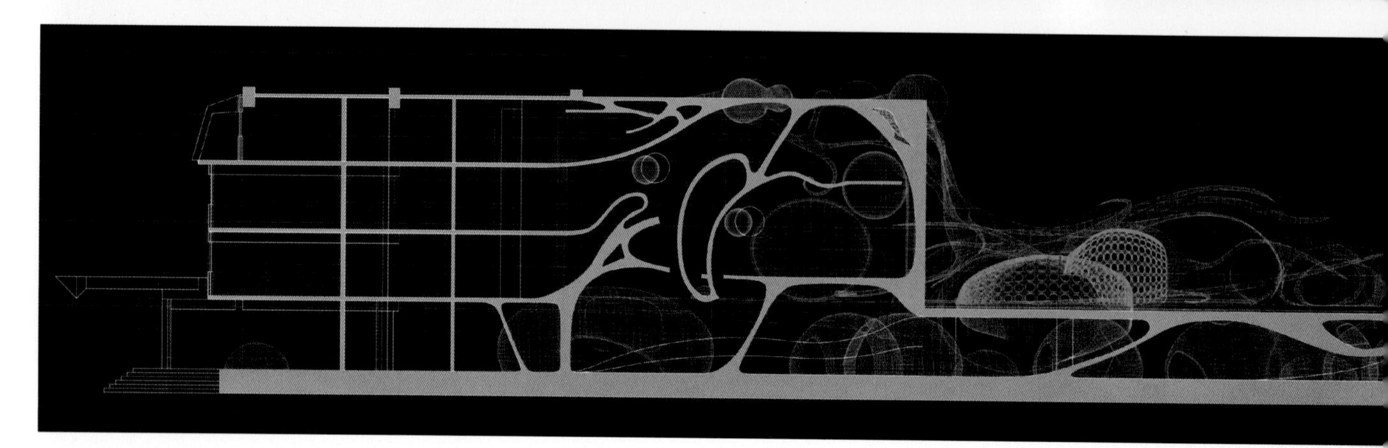

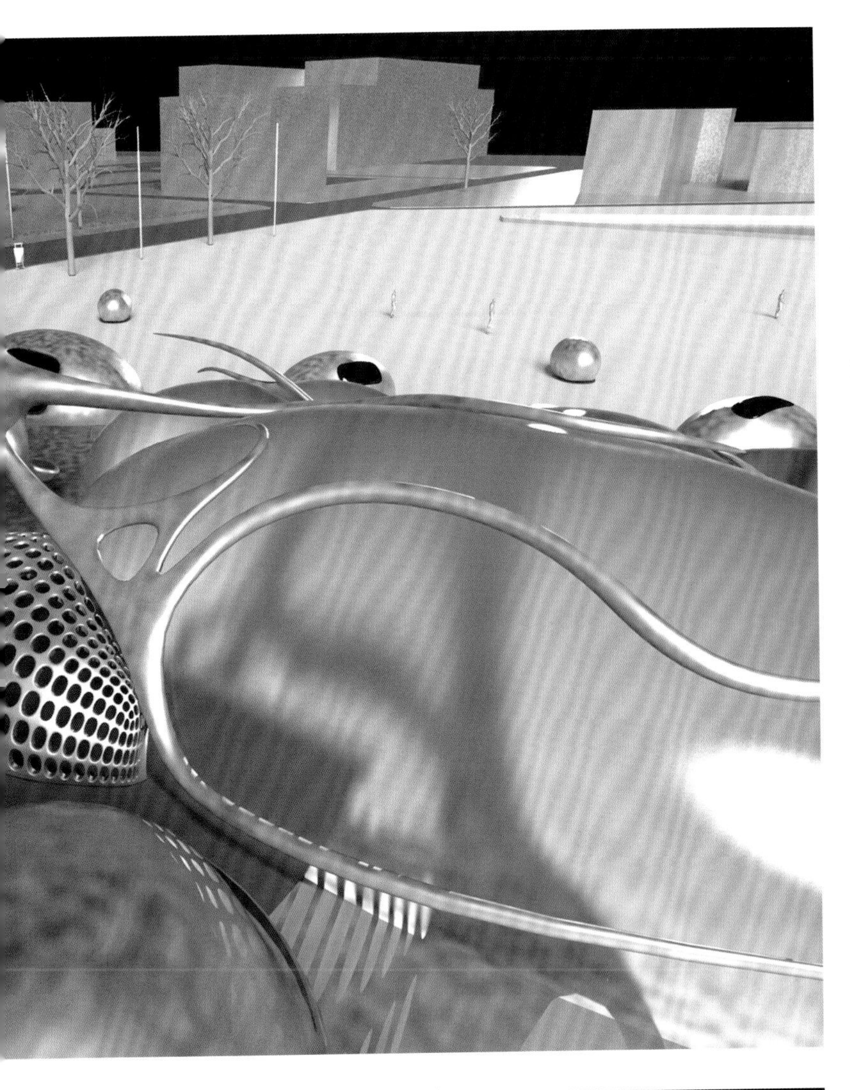

华中科技大学建筑系馆改建项目 / Department of Architecture Extension

本方案并不像一般的改扩建项目一样，在原先的建筑上加建，或者推倒原有建筑重新盖建，而是旨在体现一种“入侵”的行为。本项目并不是对原有建筑的空间、功能以及建筑能耗环保方面的重新组织，而是通过细胞的形式对学院已有系馆进行侵入，从而使人产生一种“吞噬”的感性认知。设计初始阶段，建筑师从生物细胞取得形式上的灵感，让“形式”自由地入侵“空间”，因此设计产生了六种入侵单体，同时进行吞噬。项目中，细胞这个形式迫使建筑空间产生分裂，从而出现很多小房间，可以用作工作间。在平面图和剖面图中可以看到原有建筑的墙体及楼板都被扭曲变形，但是入侵者并未完全吞噬，保留了部分遗留建筑。这样，体现了过去和未来的对比和转变。

The purpose of this project is not building and expanse on the original buildings but the act of invasion. The project is not based on an organization of function or physics or space but in the generation of sensations through cells and /bubbles that are produced to invade the stereotype of the school. At the beginning of the project, I took cells in biology for reference, and design 6 different prototypes.The space in this project are split into many small parts like "cells". As can be seen in the plan and section. The original walls start to contort and stretch to the stalk. The new specie does not completely replace the old one, but leave part of the original. So we can see the process of the infection.

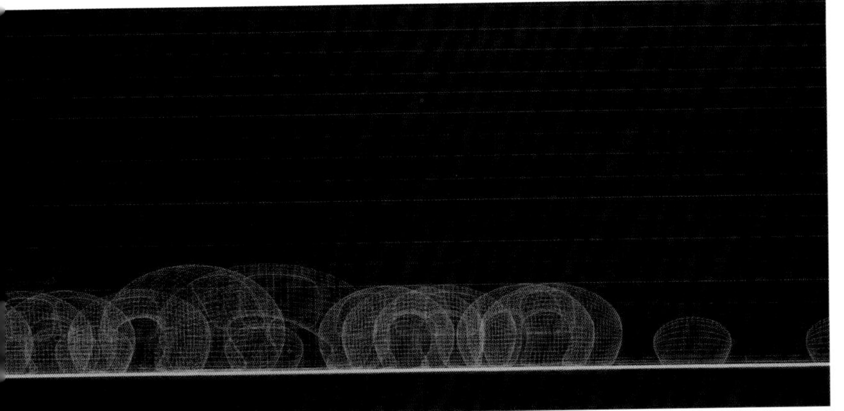

中国湖南大学建筑学院 / HNK

建筑系门厅室内改造 / Remodeling of Entrance Lobby

这是一个优胜方案，设计基于解决建筑系馆门厅受强烈的阳光直射问题。门厅的顶部由 8 个锥形玻璃采光罩组成，在长沙这样典型的夏热冬冷地区，一年有很长时间门厅都处于暴晒状态，原来这里是供师生休息与交流的区域，但几乎没有人停留；这个方案使用了日照分析与模拟软件以及 Grasshopper 插件，通过调整由 10 个连续面组成的透空折叠单元的大小与角度，形成一个连续变化的自由曲面吊顶，从而既保证了采光，有很好地解决了日照过度的问题。由于事先在 Rhino 模型中已将各部分精确定位，所以最终成果的构件尺寸非常精确。通过从上午到太阳落山这段时间对门厅的连续观察，学生们的设计目标可以说圆满完成，这里已成为这个 20 世纪 90 年代扩建的门厅的一景，并且最重要的在于师生们现在可以在此轻松地聚集和交流。

This design aims to solve excessive solar gain in the building entrance lobby, which is intended as a communication space for students and teachers, although few really stay there as the space is overheated. The ceiling of the lobby is composed of 8 pyramids of glass, which are exposed to too much sunshine in the summer. By employing solar analysis software and Grasshopper, the new proposal sets up 10 continuous folding surfaces, which are composed by many homogeneous components, to create an undulating substrate hanging from the ceiling to reduce the solar gain while guarantee sufficient light for the lobby.

大跨空间构筑 / Large-Span Structure

设计之初期望使用膜结构，于是我们利用杆件对于拉力和压力的转化做成了一个小尺度的模型，但由于过大的受力形变而被否定了。最终通过反思利用膜的可透光性和轻质性将其作为大跨度建构的维护结构。同时在承重结构设计上借鉴了汴水虹桥的经验，加入平面和曲面上的扩展，形成一种独特的编织结构。同时我们利用这种编织的形式，为膜的附着提供了新的形式法则。在设计形态生成中，建立了一个曲面组成的形体，运用参数化的手段将形体分形为编织结构的形态并在其上加上不同形状的膜材，从而完成了大跨度建筑原型的研究与构建。

The initial idea behind this project was to build a membrane structure, but this was abandoned after a test on a small scale structure model. The idea therefore changed, and the new design brief was to create a projective skin using a membrane system. Structurally the project would be supported by adopting the system of a traditional arched bridge. The idea was that this could create a unique waving pattern while at the same time offering a supporting layer for the membrane. The final overall design ended up with a free form mass that would be covered by the membrane system.

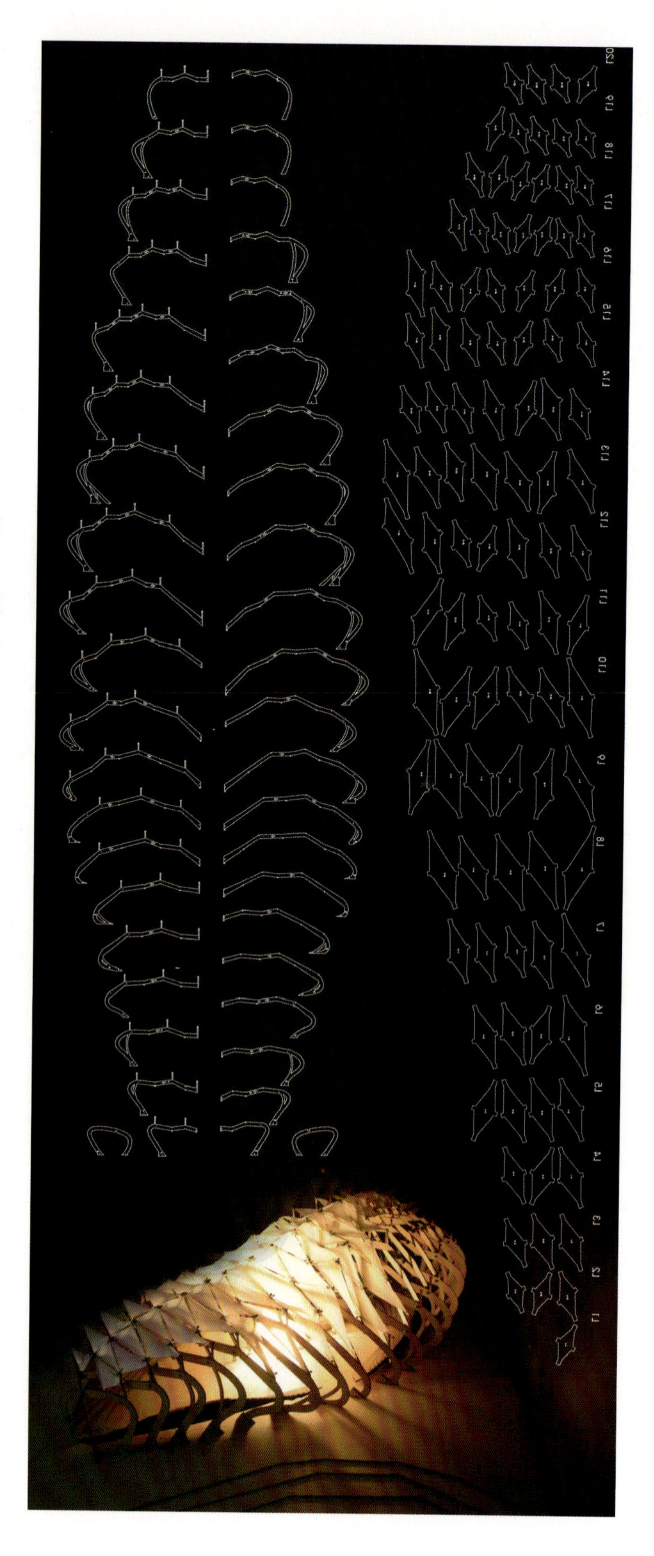

B8
A10

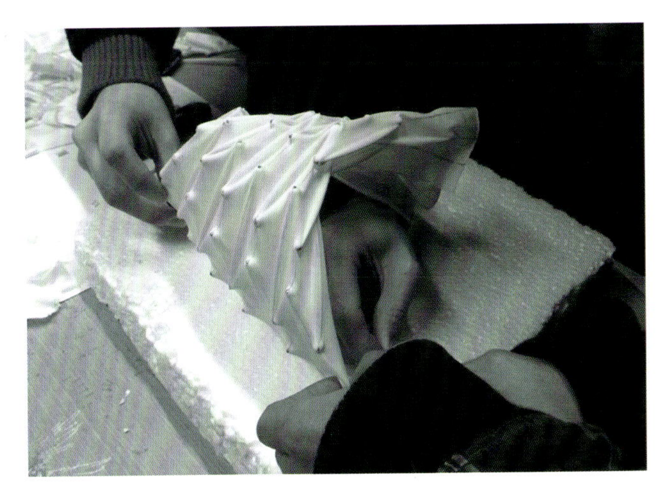

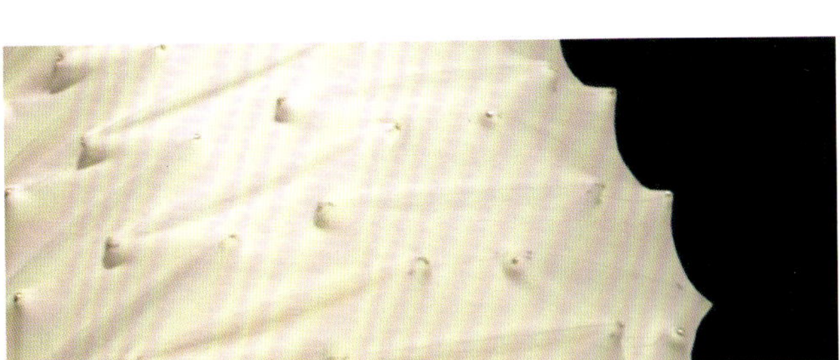

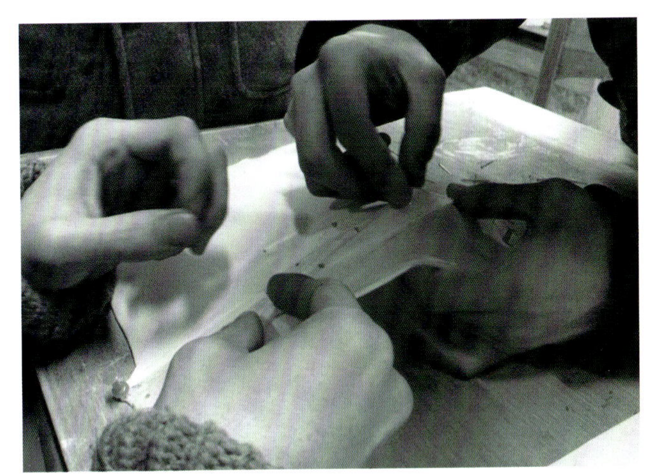

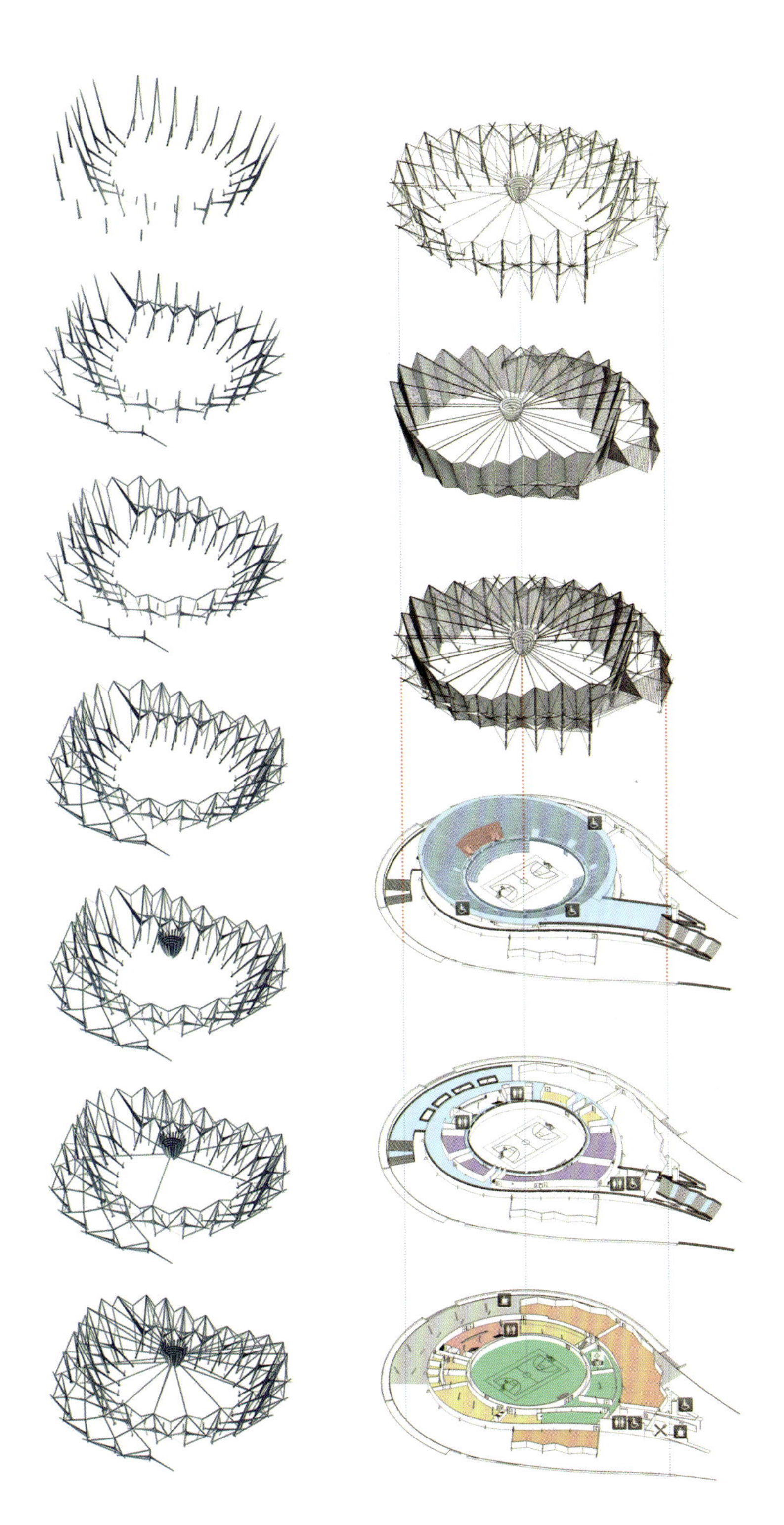

大跨度悬索结构体育馆设计 / Large-Span Space

基地位于湘江西岸，毗邻桃子湖，背靠岳麓山，是一块具有优良自然环境与浓厚历史人文内涵的地块，场所精神复杂深刻而特性鲜明。 我们充分抓住悬索的材料与力学特性进行结构设计，最大限度利用其抗拉性，结合中心的稳定环与边缘压杆构件，将屋顶荷载转化为各个单元索杆上的匀质拉力并最终导入地下，形成符合力学逻辑的稳定结构整体。最后通过同质化的手法来处理周围场地环境，使整个建筑在环境背景中和谐融入。

Located at the west bank of Xiang River, and facing the Taozi Lake and in front of the Yuelu Mountain, this site enjoys a range of outstanding geographic conditions. It also benefits from a rich historic, cultural background that also characterizes the site. The generating idea behind the design was to extend and maximize the pulling resistance of a tensile system, so as transmit all the loads from each of the tensile units to the ground, in order to create a stable structural system. The same strategy is then also deployed to create the environment surrounding the building.

中国华南理工大学建筑学院 / SCUT

数字斯卡帕 / Digital Scarpa

这个项目中，我们通过重建 Rabattement 透视法去重建一个室内的角落。这种方式帮助我们将空间折面旋转并投影到平面上。数字化的工具简化了这一过程，却掩盖了基本操作的原理。虽然 Rabattement 透视法比较繁琐，却重新清晰阐释了二维与三维转换的过程。

In this project, there was an attempt to reconstruct a corner of an interior space using computational tools. In particular rabattement perspective apparatus was employed. This allowed the surface to be rotated and then projected onto the ground plan. Other digital tools were used to facilitate this process, but their fundamental working principles were concealed. Although the process of using the rabattement projective apparatus is laborious, it manages to recreate the articulation of the transformation between 2 - dimensional surfaces and 3 - dimensional space.

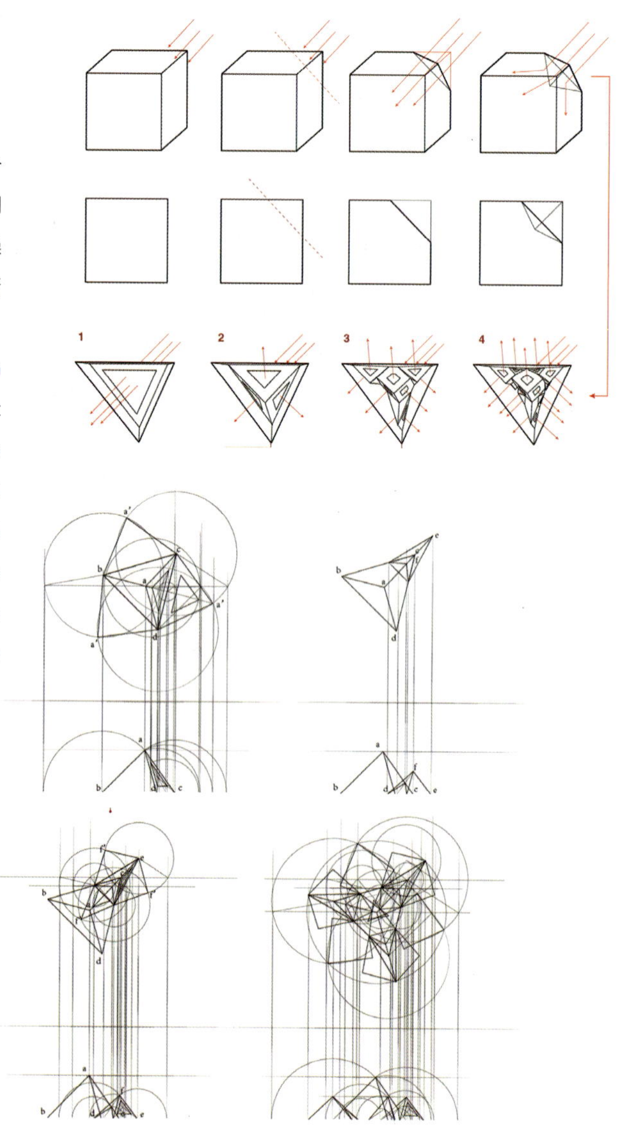

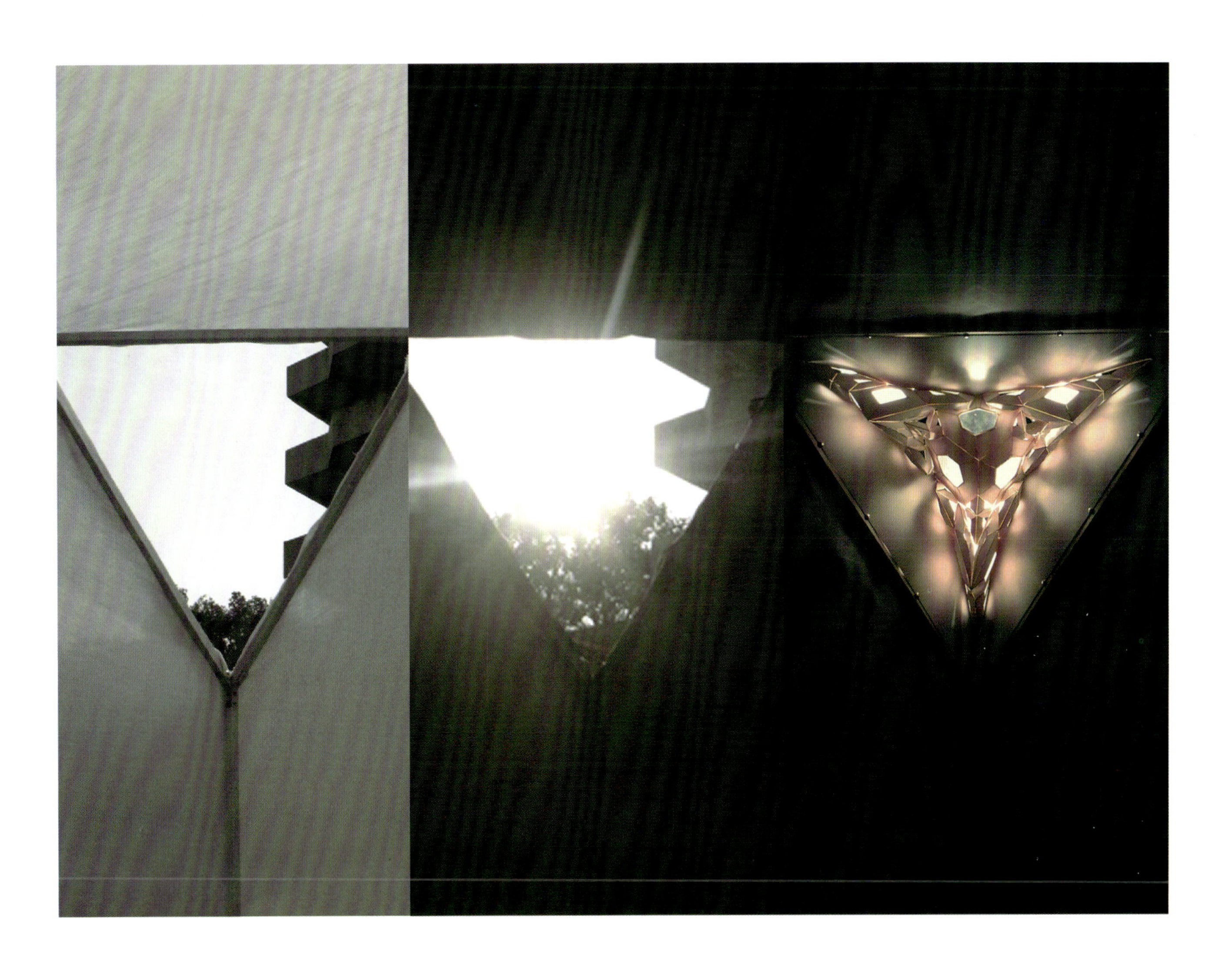

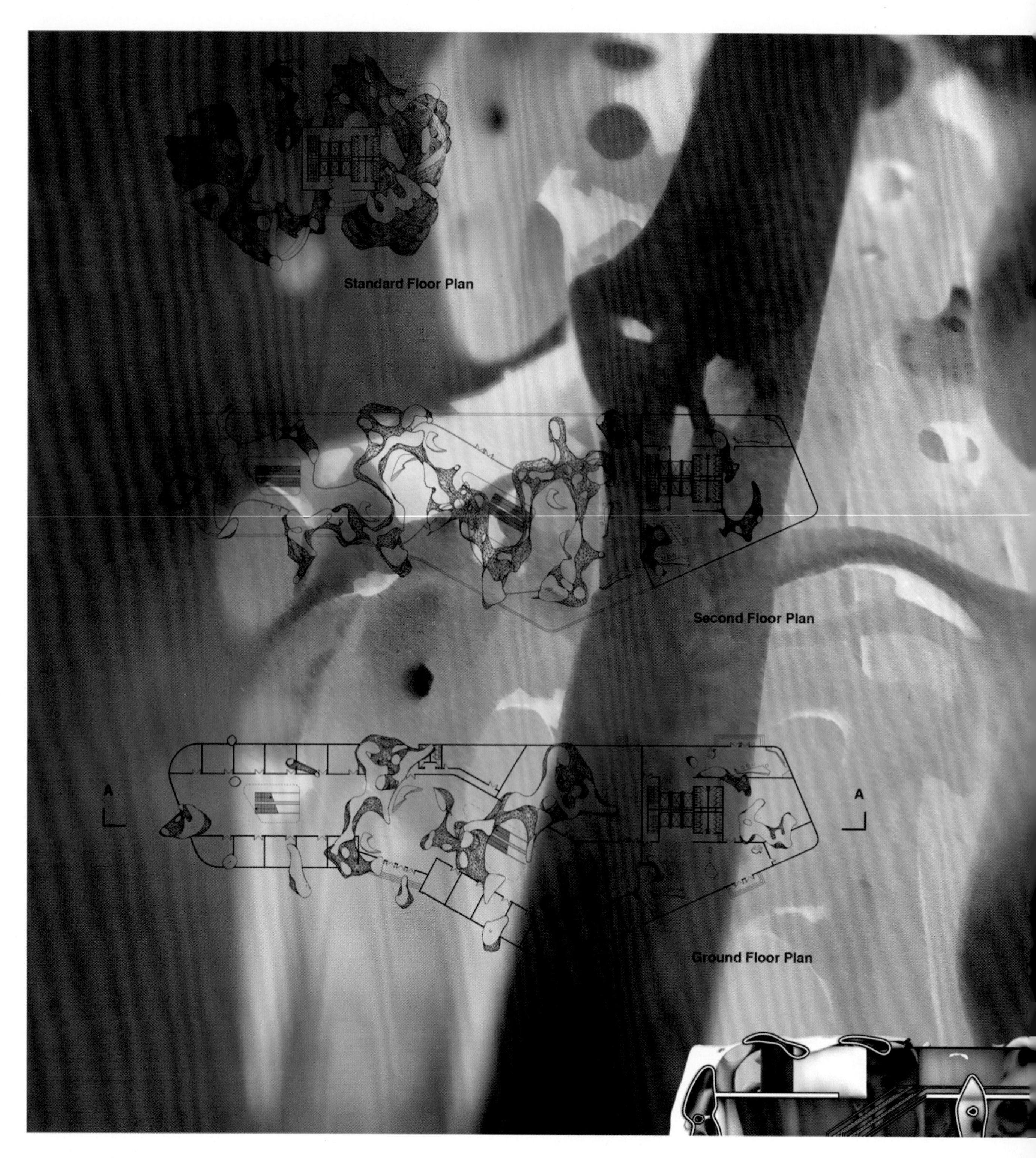
Standard Floor Plan
Second Floor Plan
A
A
Ground Floor Plan

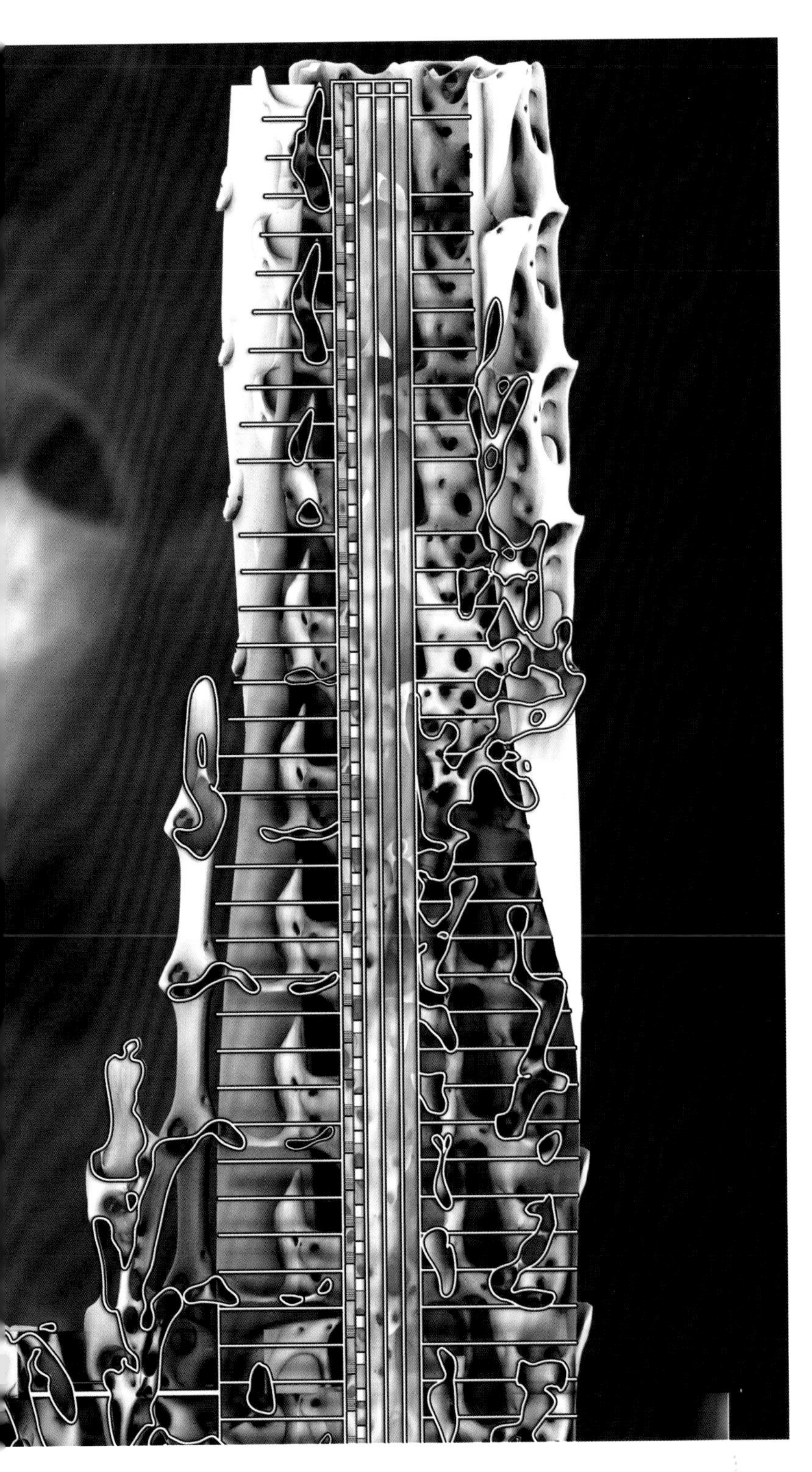

生成高层 / Generated High - Rise

这个设计课程的关注点在于以规则体系为基础的高层设计。从对自然界的动植物的行为研究中，发现其中存在某些内在的逻辑控制着其垂直的组织和生长。这一种组织方式可以被抽象成最基本的点之间的关系。这一过程让我们联想到了积木游戏。在设计中，通过定义自己的游戏规则去创造一个开放的系统，而不是传统的自上而下的方式，并通过这个系统去控制积木的摆放。

This studio is focused on developing new rule - based strategies for the design of high rise buildings. Based on initial research into natural performance, it was found that there are some latent relationships that control the generation of vertical form. This kind of field condition could be abstracted into the most elementary organization of points. This in turn inspired the project to develop into a form of game, the Lofty Tower Games. Rather than following the universal top - down strategy, an attempt was made to create an open system by defining the rules to play with the Lofty Tower Games in voxels, as the basic geometries.

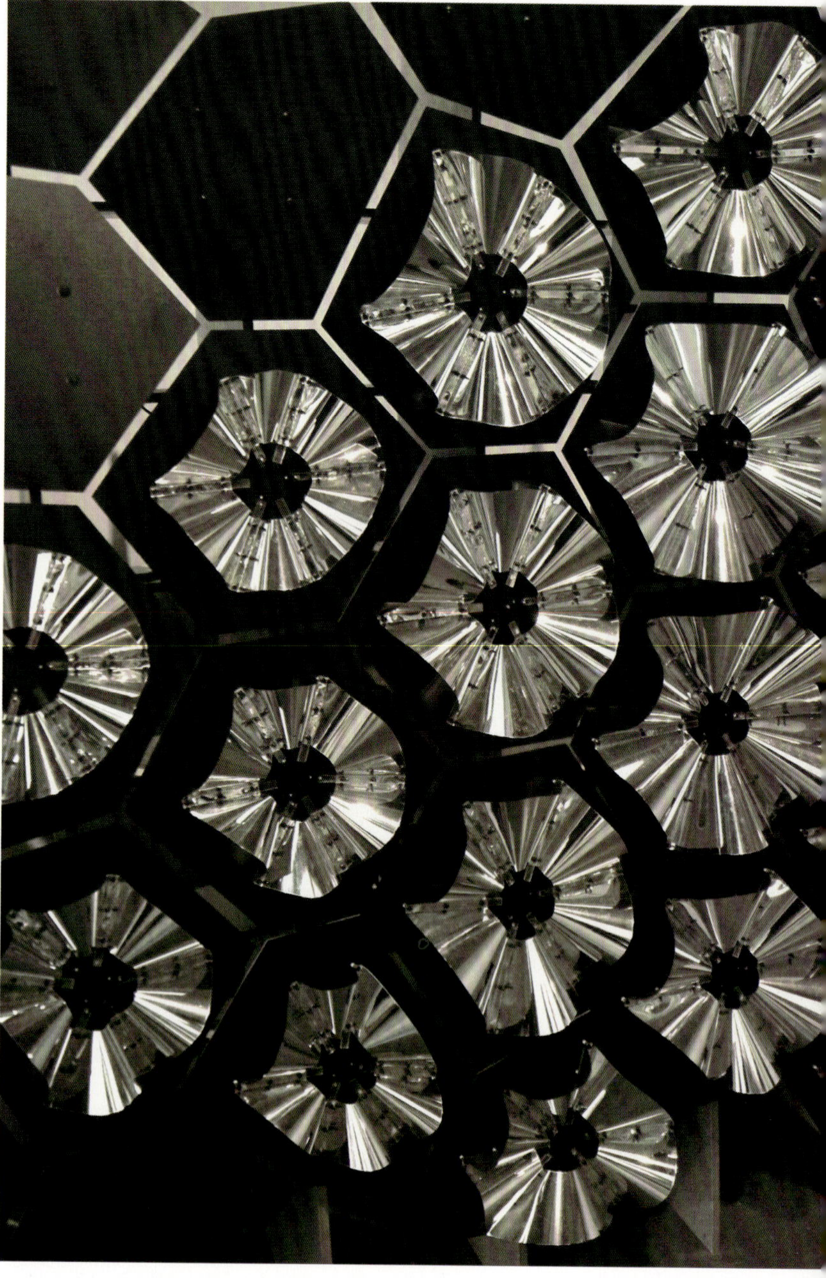

热感墙 / Thermo Surface

我们选取 Felix Candela 的双曲面结构进行研究，理解了几何形式美与力学特性相结合。然后我们从他的固有形式中发展出可互动的形式。互动单元由两个舵机和一个夹在中间的曲面构成，每个舵机接收来自单片机的信号而改变角度。由此，我们进行设计，选取建筑内墙面将其分割成多边形面板，每个面板上附着可变单元。通过单元的变化与运动，使建筑的墙面拥有智慧并根据认知需求改变形态，使建筑更加生态可持续。

After researching Felix Candela and his use of the hyperbolic paraboloid, an attempt was made to develop his static geometry into an interactive one. The interactive component consists of two servos and a surface between them. Both servos can receive signals from the microcontroller and then stretch or bend the surface, changing their apertures based on the air volume and wind direction. Based on this, a design was generated. The building interface was divided into multiple voronoi panels on which several interactive components were placed. Through movement of this component system, the building interfaces have an intelligence to create a more delightful interior environment, making the building more sustainable.

中国天津大学建筑学院 / Tianjin University

回转 / Revolve

本方案是内蒙古五当召的藏传佛教博物馆。它是收集、展览和学习藏传佛教教义的场所。设计选址在山脚下，西面朝向寺庙入口，有两道泉水从基地的东西两侧流过。设计者从几何和图形分析开始概念设计，选用双层莫比乌斯环为原型，将流线和功能组织在一个扭曲的独石形态中。设计以纯粹的形式展现一种对宗教和神圣感的哲学思考。

This is a design for a Tibetan Buddhism Museum at Wu Dang Lamasery in Inner Mongolia. The museum is expected to be a remarkable place to collect, exhibit, and study all the relics and data about Tibetan Buddhism. The project is located at the foot of the hill, facing the west of the entrance square of the Lamasery, with two streams flowing along the east and west sides. The design started with a geometric and graphic analysis. Composed of two Mobius circles, it incorporates the circulation and functional elements within a twisted monolithic form. The pure appearance of this design was intended to evoke a philosophic sense of religion.

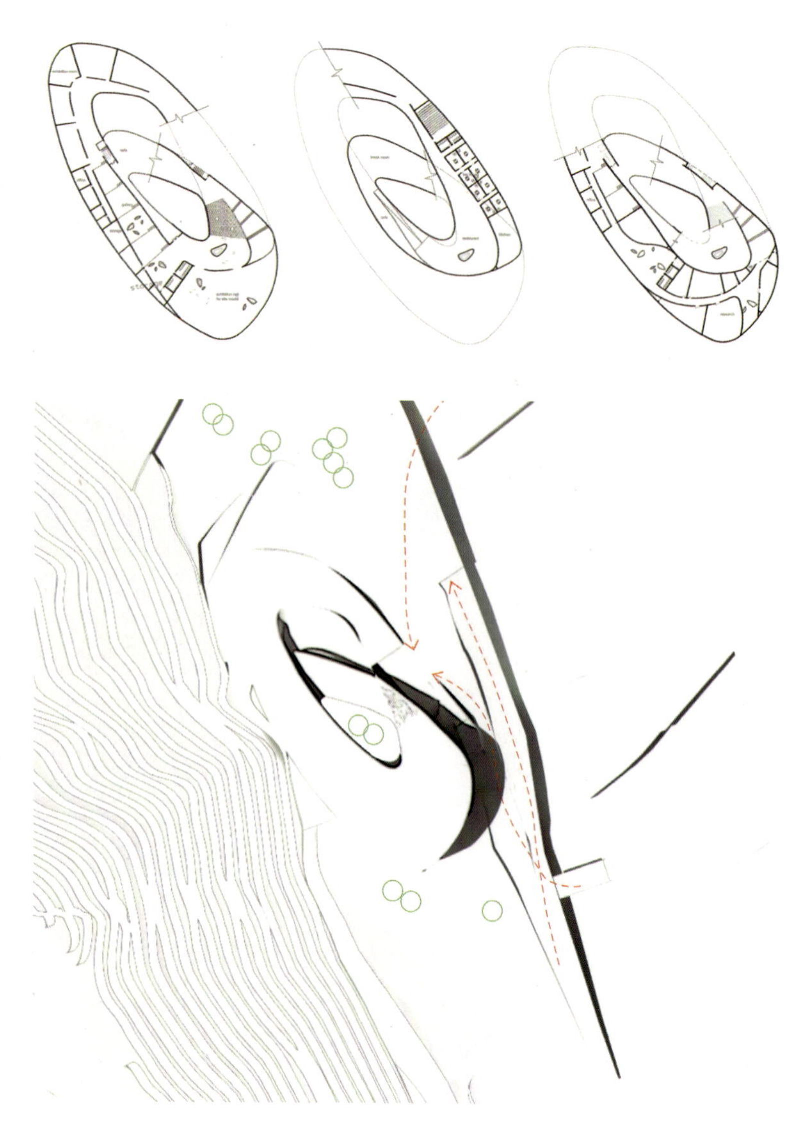

神圣之光 / Holy Light

该设计为内蒙古自治区的五当召藏传佛教博物馆。设计者将建筑创造性地分为几个简单的几何图形，每个图形都代表着不同的空间类型。室内设计中光线运用非常出色，巧妙地刻画出空间的形状。参观流线被光影所限定并指引，完成了空间的叙事结构。景观设计也强化了藏传佛教的氛围，阳光随时间变化而变化。

This project is a design for a Tibetan Buddhism Museum at Wu Dang Lamasery in Inner Mongolia. The designer creatively split the building into some simple geometric shapes that represent different typologies of space. The interior light is well - designed to be a key element in shaping the space. Circulation is defined and traced by illusions of the light and shadow, the arrangement makes space a unique narrative construction. The building makes an effort to keep in harmony with the mountain, stream and environment. The landscape design also enhances the atmosphere of Lamasery, as well as the changing sense of sunlight.

运河方舟 / Canal Ark

设计坐落在天津杨柳青的元宝岛，大运河流经整个基地。这个设计是一个包括居住社区、零售店、博物馆和公园的建筑综合体。设计者从流线研究开始，利用 Processing 软件分析步行和自行车流的特性，生成基本的路网系统，然后利用犀牛和参数化工具刻画木结构的细部。景观部分在折叠之后自然形成公共空间和体量，景观下面布置了零售和公共活动等场所，景观上一层位置安排了居住单元等功能。

This project is located on Yuanbaodao Island, where the Grand Canal flows through the region of Tianjin. It is a design for a building complex including a residential community, retail spaces, a museum and a park. The design started with an analysis of walking and cycling patterns to generate a basic road system using Processing, and then Rhino and Grasshopper were used to make a more detailed and precise model. The landscape was carefully folded to give the design a dynamic sense of space and mass, with retail spaces and public spaces underground connecting with the neighborhood community, and apartment units arranged on the ground floor.

中国同济大学建筑学院 /TJU

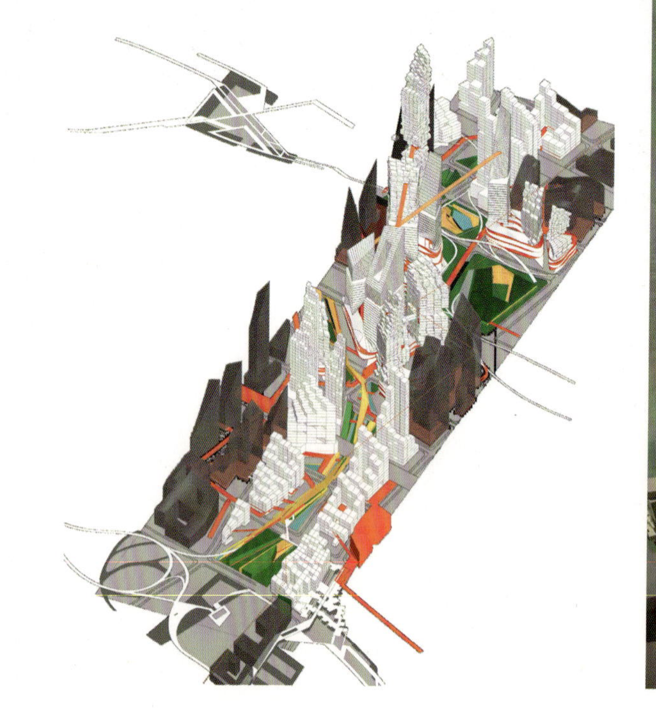

微观城市 / Micro Cities

微观城市使地区二分法寻找到了共同点。在那里，数量满足了质量，单体的个人产生了集体的性格，传统激励了现代性，技术补充了自然。这是一座拥有身份和强度的城市，IDenCity，是个集合的有机体，国际性的价值在其中收获。高品质和高性能的生活在对于自然的最大程度的接近中、对于传统的赞美和技术的热情中而达到。

Micro cities are cities where dichotomies negotiate common ground. Here quantity meets quality, singular personality produces a form of collective character, tradition inspires modernity and technology complements nature. This is a proposal for the micro city of IDenCity. It is a city of identity and intensity, a collective organism whose global values are being harvested from within. In IDenCity high quality and high performance living conditions are achieved throughout by the close proximity with nature, by an appreciation for tradition and by an enthusiasm for new technologies.

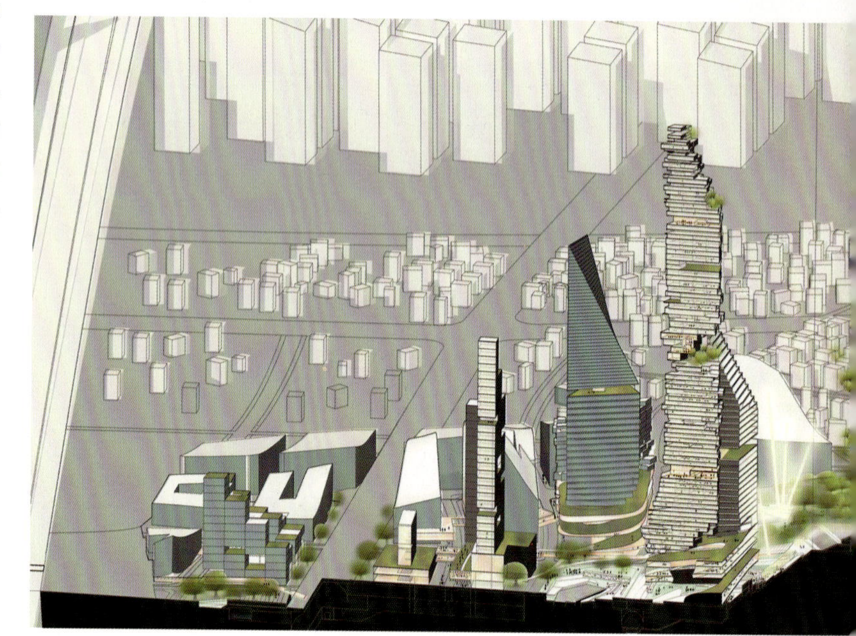

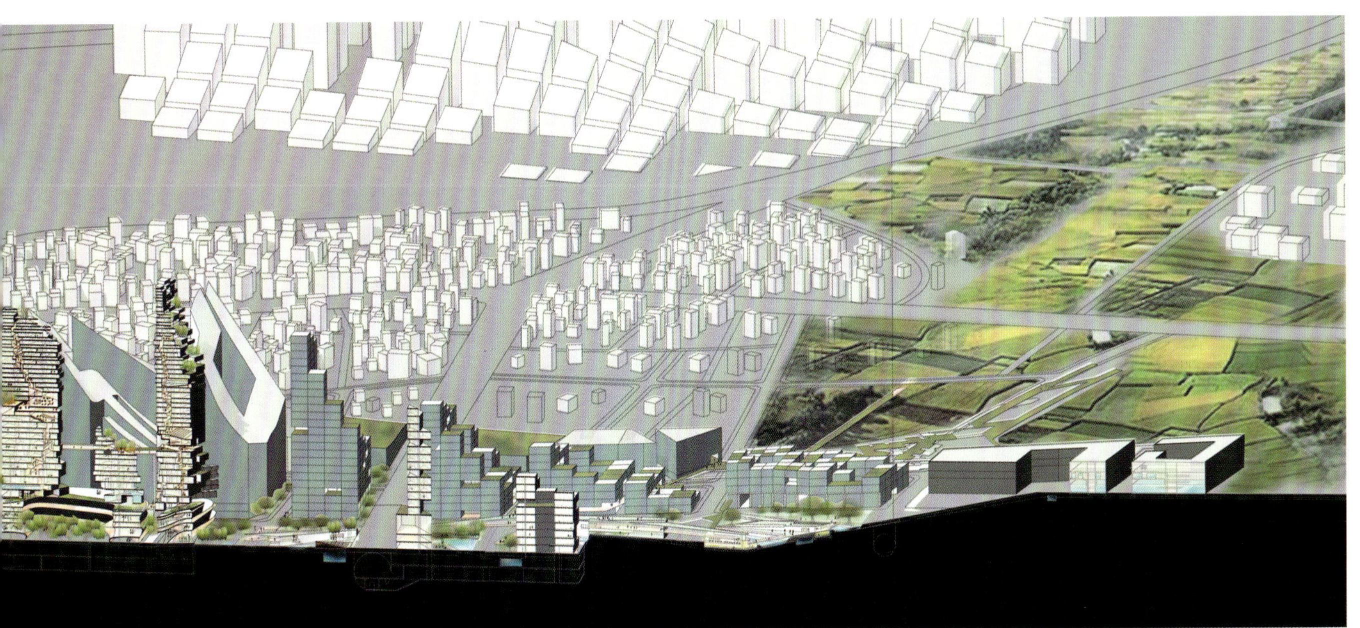

2012 欧洲太阳能竞赛 / Solar Decathlon Europe 2012

“复合生态屋”将主动式和被动式的环境系统共生地结合进彼此之中，两种设计哲学融合并在互动中彼此受益。同时参数化的设计方法和生态策略也结合到建筑设计的语言之中，满足功能和生态的需求，创造了低碳生活的未来图景。

The "Para Eco-House" is a design for a low energy house for a low carbon future that was submitted as an entry for the Solar Decathlon Europe 2012. The Para Eco-House combines both computational and ecological strategies within the logic of the architectural language that is used in the design of the house. Both 'passive' and 'active' energy systems are utilized in this project. In going beyond the functional and environmental requirements, the project attempts to create a paradigm for the design of a house for a low carbon future.

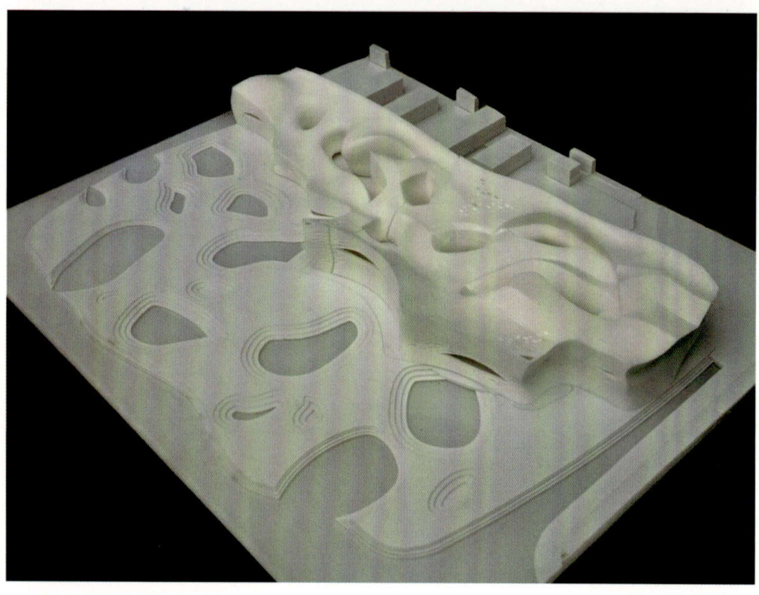

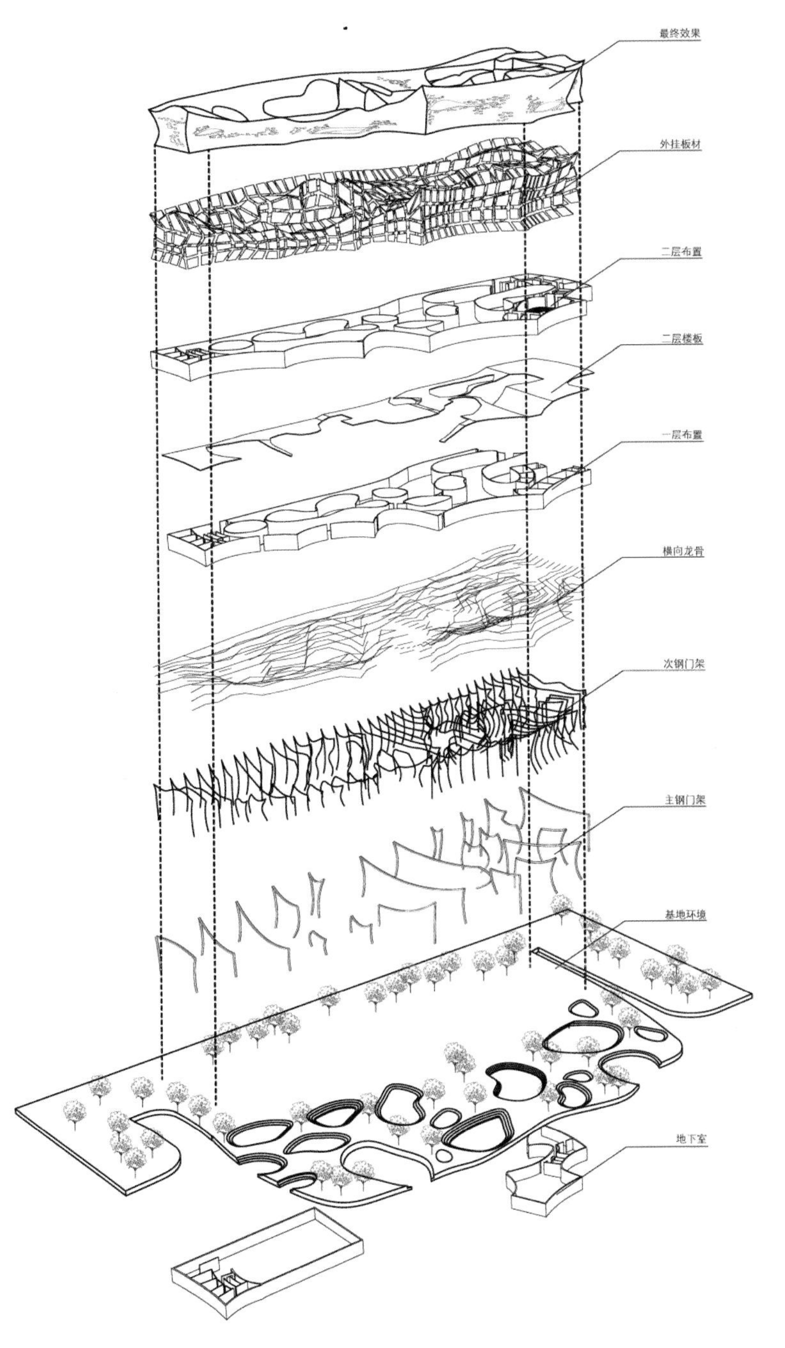

西溪湿地新西泠印社博物馆 / Xixi Wetland Museum

在本次教学实践中，形态发生学的生成算法主要体现在算法思路的建立、代码编程制作工具包、建筑功能参数与环境参数的设定、参数与工具包结合等过程中。本设计通过模拟西溪湿地的肌理发生机制，并以此为工具，受到建筑本身参数的影响，作为建筑生成的逻辑并最终生成建筑。

This project was part of the 2012 Chinese Eight-Universities Joint Design Workshop. During the teaching practice, the use of the algorithm of morphogenesis is mainly reflected in the general algorithm ideas for code, programming tool kit, building and environment parameters setting and the integration. Here we used the simulation of the texture generating in Xixi Wetland as a design tool in order to generate the form of the architecture. In this project this strategy can therefore also be seen as the basis for the logic of the architecture itself.

中国清华大学建筑学院 / THU

U 胶 / Glue

我们生活的现实世界是由各种复杂系统组成的。通过观察现实世界的物质运动，可以开始理解各种不同的集群行为。几滴 U 胶被两块玻璃板压平，然后在胶凝固后再将玻璃板拉开，U 胶便收缩成为枝杈状图形。接下来，使用 Grasshopper 编程模拟了这一现象中 U 胶粒子的运动和聚集，以及孔隙的发展，并引入不同的参数来影响这一过程。最终的算法被用于展览馆的设计，生成了尺度亲切怡人的室内空间以及室外小径和院落。

The physical world is composed of various complex systems. Through observing the motion of matters in the physical world we can start to understand various swarm behaviors. Two glass panels were laminated with drops of glue. When pulled apart before fully congealed, we could witness the semi congealed strands in between. The algorithmic tool Grasshopper was used to simulate the movement of glue particles, and the development of the gaps. Different parameters are tested to control the process. Finally a 3d version of this algorithm was used to generate a gallery design with intimate outdoor alleys and courtyards.

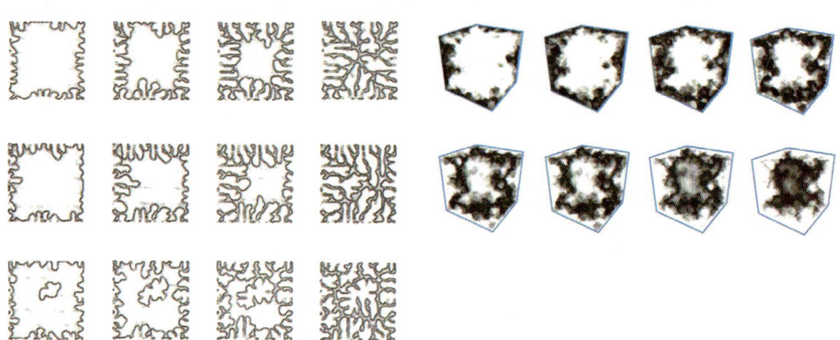

限制性扩散聚集算法 / DLA

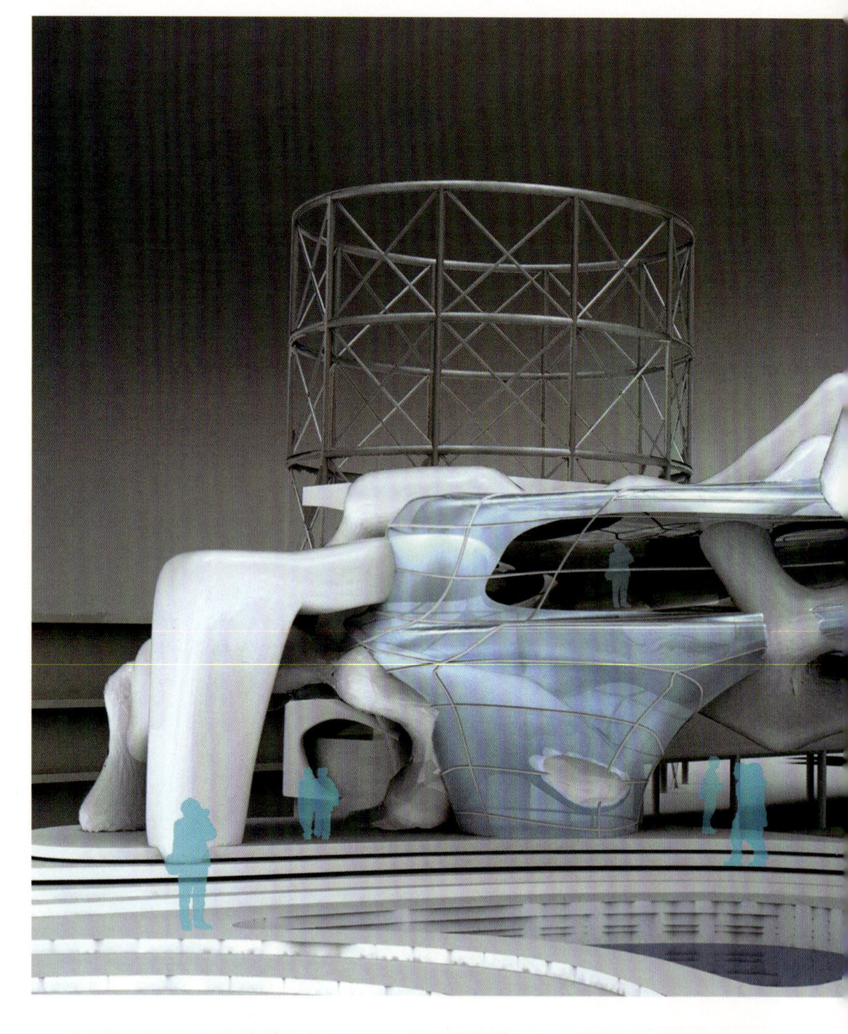

人在展览空间中的行为可以被看做是没有特定目的的闲逛。基于这样一种认识，限制性扩散聚集算法（DLA）被用来模拟展览空间中的人流。这个设计是位于工业遗迹区域中的一个展廊。首先通过对场地的视线分析，确定了建筑体量中需要避开的，让人的视线能够看透的空间。接下来用 DLA 算法生成流线系统。这一算法生成的枝杈系统能够避开视线的空间，并能够相互连接形成网络，然后使用变形球技术包裹流线。最终的设计是由虚实两个体量组合而成，实体是结构和展室，虚体形成公共空间。

People's movement inside a gallery space is considered random. Based on this understanding, the algorithm of diffusion limited aggregation (DLA) could be employed to simulate this kind of circulation. The design is for an art gallery located in former industrial park. Sightline studies define some restricted spaces in the building volume that will allow people to look through. Then the DLA algorithm is developed to allow random growth of branches that will form an interlinking network, and yet avoid stepping into the sightlines. Metaballs were used to envelop the circulation. The final design has solid volumes for the structure and galleries, and transparent volumes for the public space.

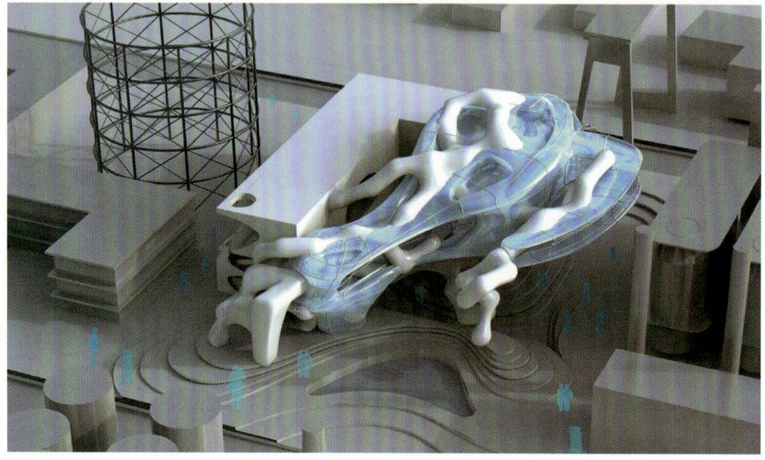

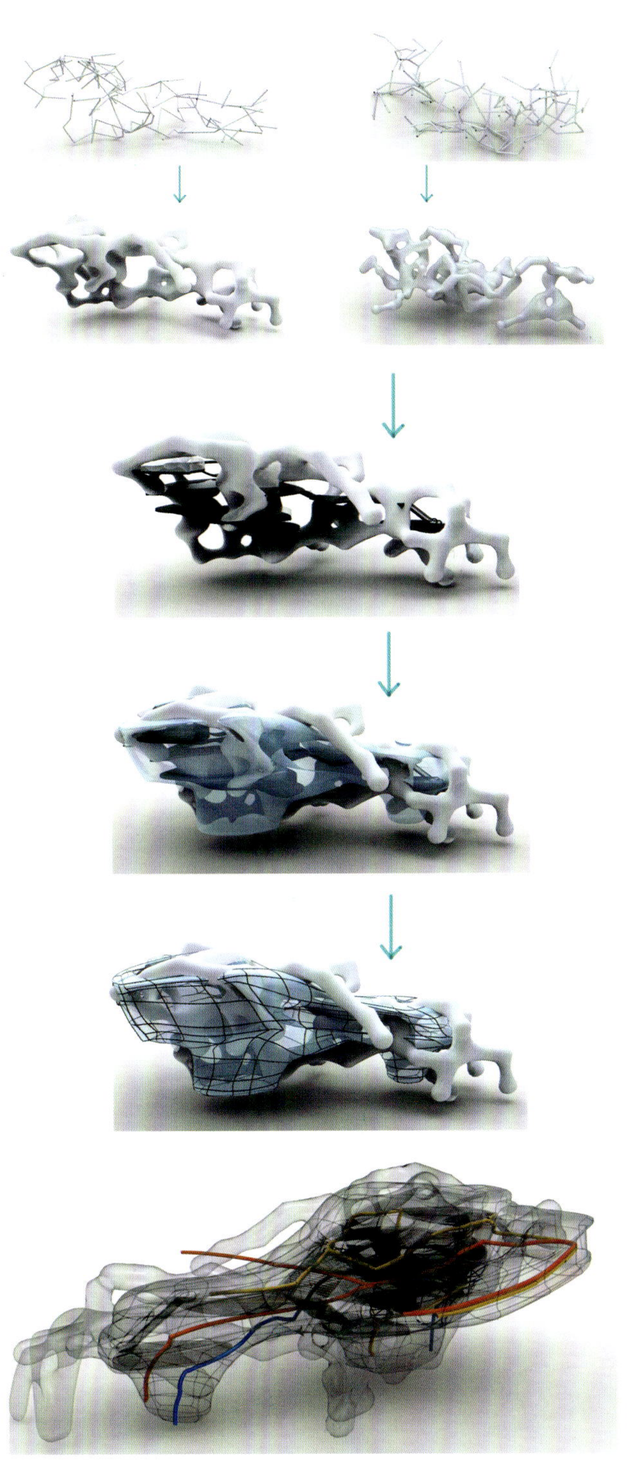

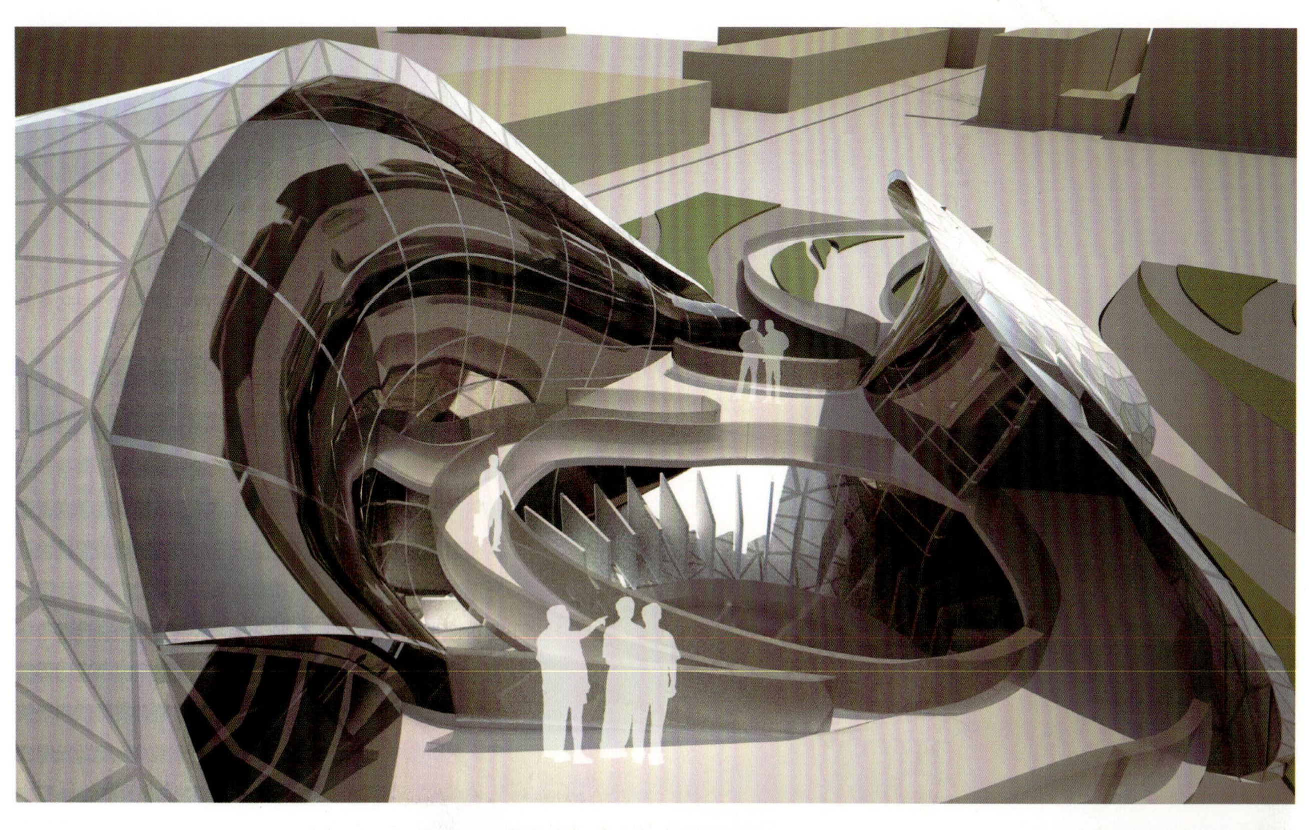

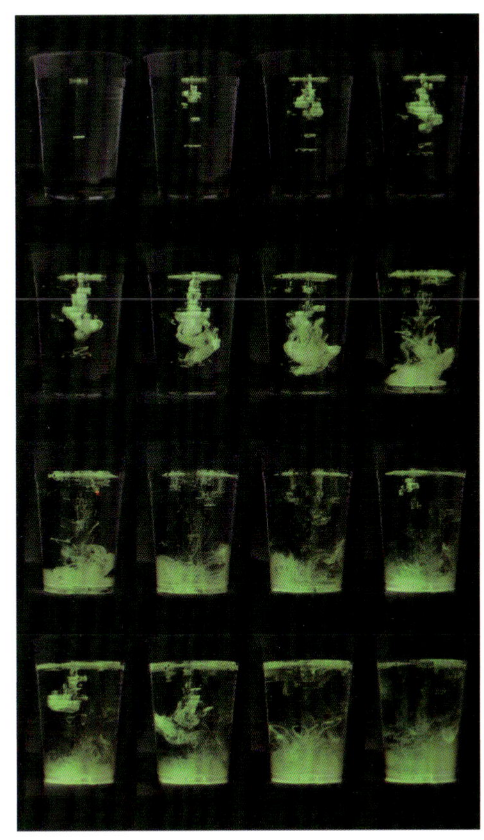

混沌扩散多项式曲面体 / Chaotic Polynomial Curves

荧光液滴的扩散属于混沌运动。为了模拟这一运动所展现的复杂形态，用 Chaoscope 软件的多项式算法近似地进行了模拟。这一软件生成的形体是点云，先将图形点云导进 Matlab，再从 Matlab 通过程序导进 Geomagic，让点构筑 Mesh，优化 Mesh 后，将 Mesh 导入 Rhino 再进行编辑，最终得到设计形体。在北京 798 的一街角计划建设一个艺术画廊，其建筑形体基于场地的条件及使用的要求，使用这一数字图解工具生成设计方案。这一工具的特点在于能够生成连续、流动的混沌型曲面体。

The diffusion of a fluorescent droplet is a chaotic movement. The polynomial algorithm in Chaoscope software was selected to approximately simulate the complex morphology of that movement. By rewriting the formulas in Chaoscope, corresponding point-cloud was generated in Matlab. The point-cloud was then imported into Geomagic to build-up mesh. After optimization, the mesh was again imported into Rhino for further editing and finalization of the form. The above-mentioned procedural tool was used to create an art gallery located in the corner of Beijing 798 historical zone, where the design of architectural form was based on the site conditions and the requirements of the program. This tool has the ability to generate continuous, chaotic curvature forms.

中国西安建筑科技大学建筑学院 / XAUAT

基于拓扑优化的建筑设计 / Topological Optimization

此次设计的主题是研究基于 Inspired 软件的拓扑优化结构计算，既然是拓扑优化，其载体是整体结构，手法是拓扑变形。既然设计的命题是全新的，其新结构形式的出现必然解决了现阶段一些无法解决的空间问题，方案以此为出发点，通过研究现有办公楼的优劣势，进行空间设计方法的创新，试图通过简单的尝试，解决一些传统框架结构无法解决的空间问题，完成命题的尝试。通过简单的研究，发现传统办公建筑中，办公空间的功能单一、环境枯燥，容易对员工产生负面影响，所以设计着重解决各层的空间联系，通过引入“交通岛”的概念，解决目前的问题，在形式的选择上，选取了均质的圆形，通过四个偏心圆的办公体块进行拼接，达到了空间上的连续，在交通空间上升的同时，办公空间等功能空间也同时连续上升。并且在屋顶上形成了连续上升的屋顶平台，将交通岛的设计概念进行延伸。

This design is for an office space using structural calculation of topological optimization based on Inspired software, with the overall structure as the medium and topological deformation as the technique. Finding a number of problems with existing office layouts, the focus of the design was to deal with spatial connection among the between each level by introducing the concept of the 'traffic island'. A uniform circle was used as the connecting form feeding four circular office spaces to generate a continuous space, with the offices rising alongside the traffic space. Finally a continuously rising roof terrace was formed to extend the 'traffic island' concept.

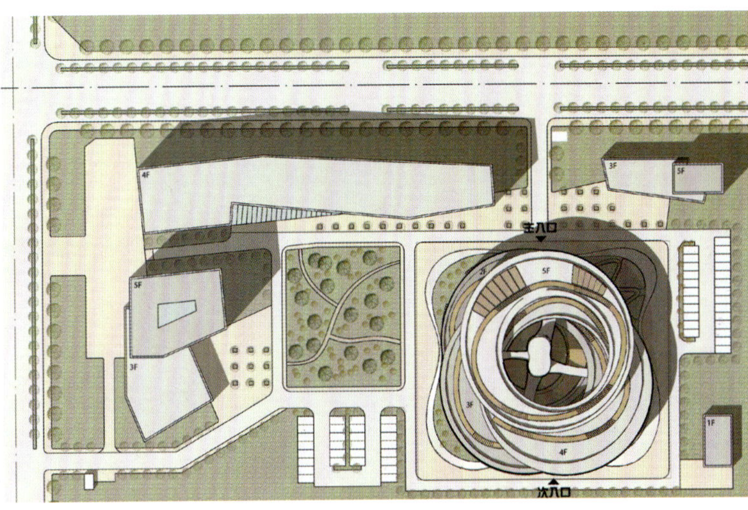
4F
3F
5F
主入口
5F
2F
5F
3F
3F
4F
1F
次入口

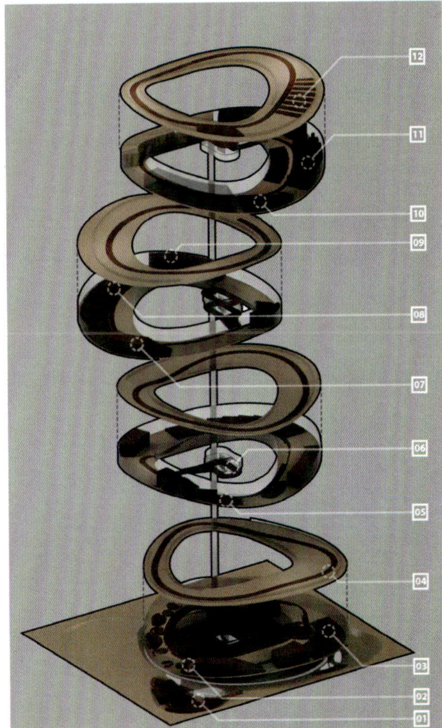
12
11
10
09
08
07
06
05
04
03
02
01

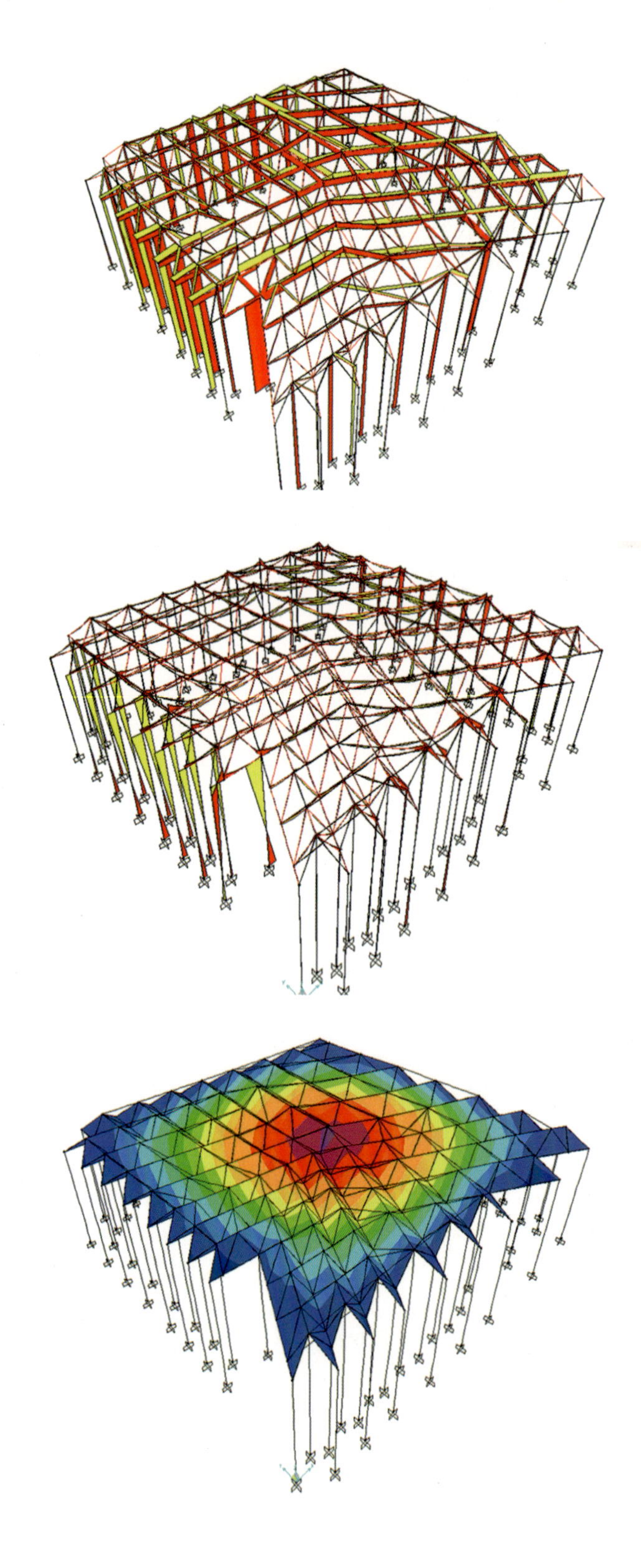

现代苹果产业信息技术服务中心 / Information and Technology Center

在设计过程中，对折叠的特性，折叠的数理描述进行研究，之后再在 Grasshopper 的环境中对其进行模拟生成程序编写，确定可控因素，从而达到对它的控制。在具体的方案设计时，选出相应数字化的折叠语言，在主体形态的产生过程中作出了对西北地貌特征的回应，模拟的形态体现的是一种台地的特征，同时又突出建筑的标志性。同时利用数字化可控参数的调节，对应具体空间的大小、形态的生成。 整个设计过程，用计算机模拟折叠，对折叠有一种精确性的控制，这时再来用这种有参数控制的折叠语汇回应建筑的问题，最终达到的结果是数字化、折叠、建筑设计之间的交互。

The concept of the 'fold' was studied mathematically and computationally, and a computerized program was scripted in Grasshopper to determine the controllable factors. In the process of design, the building form imitates the main features of the landscape to give the building a relationship with the local topography. Meanwhile, the overall form and each functional space were decided by adjusting the computational constraints. The whole design process was therefore generated by the computer, with the 'fold' being controlled precisely by computational constraints. Meanwhile, by being defined as a response to the building design this computationally controlled language shows an interaction between computation, the fold and architecture.

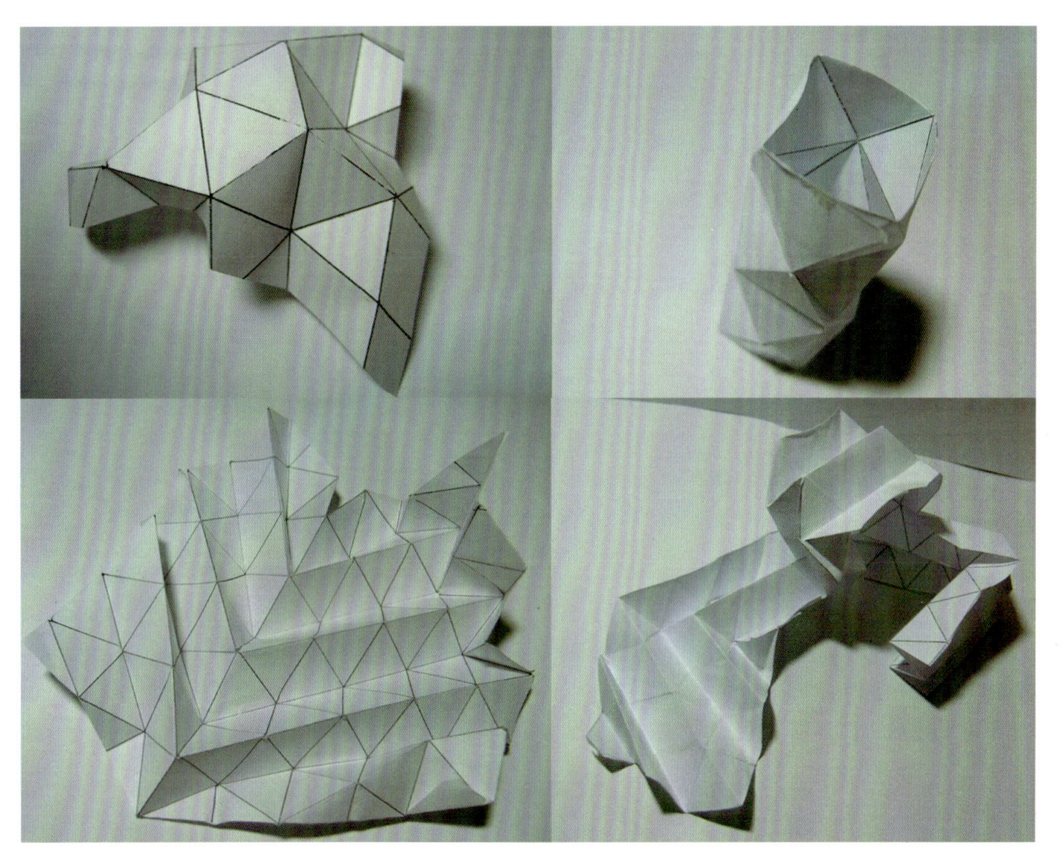

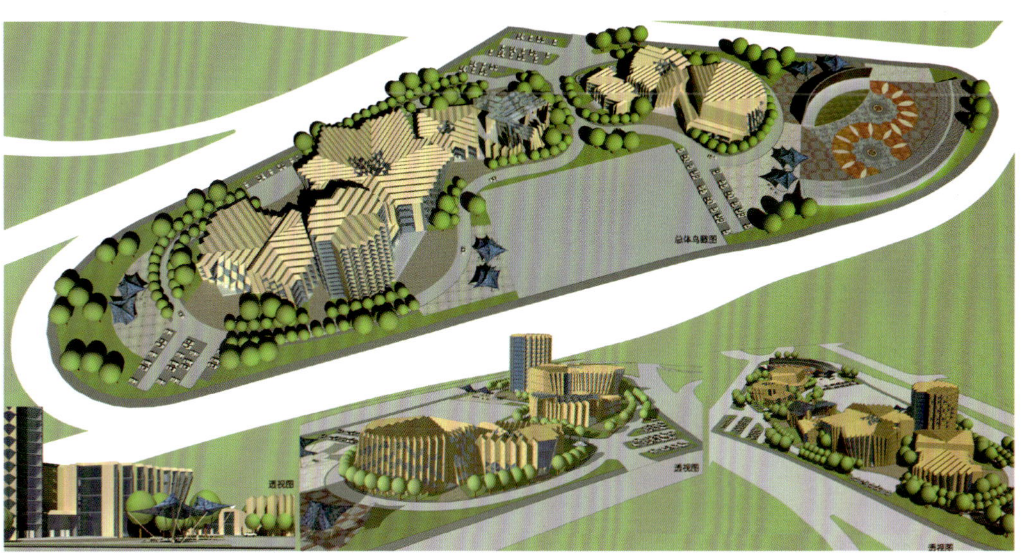
总体鸟瞰图
透视图
透视图
透视图

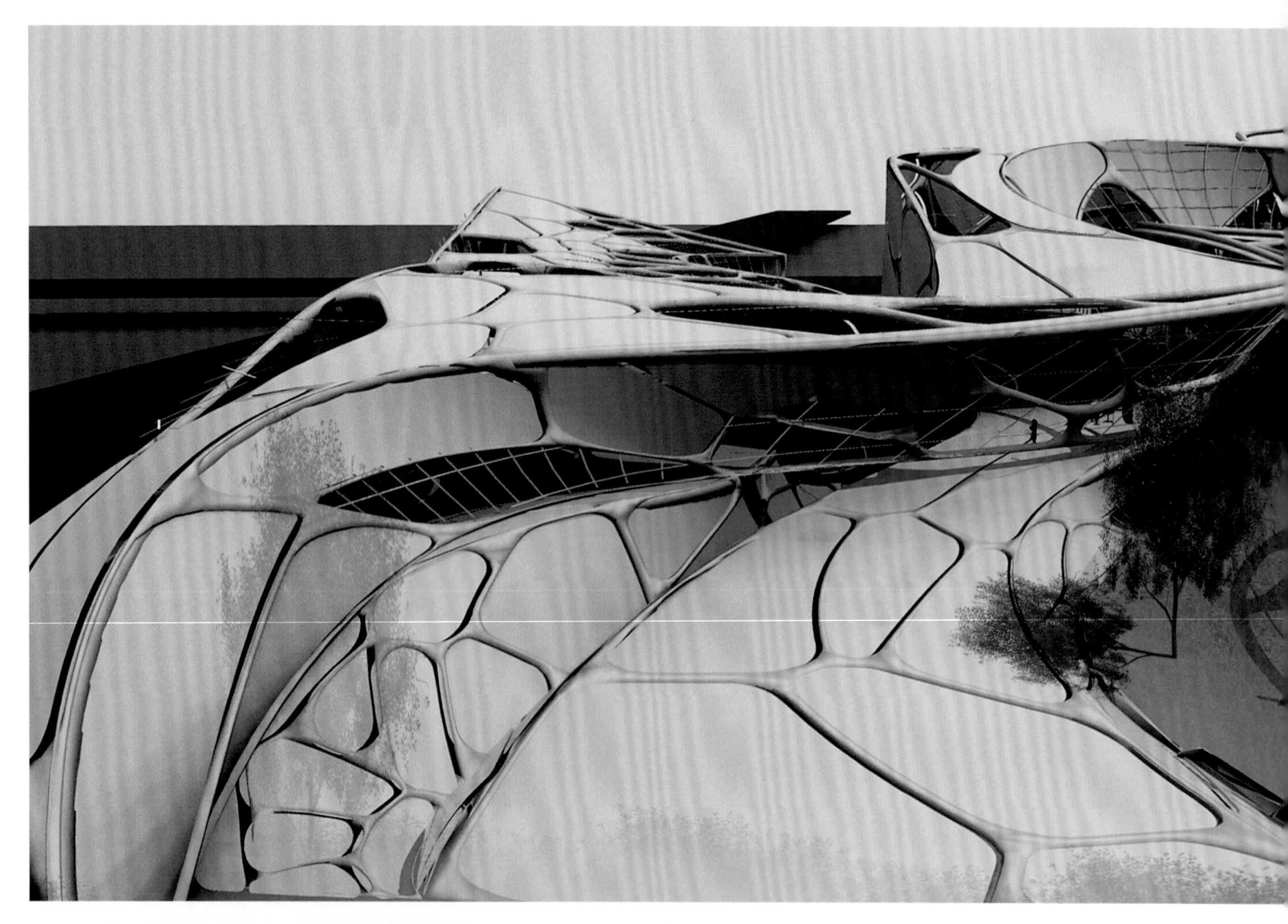

海升果业总部设计 / Haisen Fruit Headquarters Design

设计以功能与场地的拓扑关系作为原点，衍生出若干布局的可能性，针对环境、景观、场所的因素选择出一种最佳的形态，作为原始的建筑界面。借助结构有限元计算软件可以得到建筑曲面的受力分布图，引入 voronoi 的理念，对应结构计算出来的受力云图进行分布控制点，形成相应的结构体系，同时作为建筑表皮的元素，结合功能布局设置墙体、玻璃、模糊了内部与外部的空间界面，使用空间与灰空间相互穿插，形成一个柔和的共生体。

The project is based on the topological relationship between function and site, which generated several possible layouts from which the most suitable shape was selected. As a first step the force distribution of the architectural surfaces was analysed through structural finite element analysis software. Then the concept of the voronoi was introduced in order to distribute some control points on the stress nephogram. As a result, a corresponded structure system was generated as well as an architectural surface. Finally, walls and glazing was introduced according to their functional distribution in order to create a compatible environment in which used space and empty space alternate.

索引 / INDEX

America

Columbia GSAP

Narcissus & Nemesis

Andriana Maria Koutalianou,

Luis Felipe Paris

Tutored by Ezio Blasetti

Brain Hacking

Matthew Celmer

Tutored by Toru Hasegawa, Mark Collins

No 'Heimat' for You

Pablo Costa Fraiz, Matt Miller

Tutored by François Roche, Ezio Blasetti,

Farzin Lotfi-Jam.

Harvard GSD

Horizon House

Carlos Cerezo Davila, Matthew Conway, Robert Daurio,

Ana Garcia Puyol, Mariano Gomez, Natsuma Imai,

Takuya Iwamura, Thomas Sherman

Tutored by Kiel Moe, Mark Mulligan

Becoming-Other

Mariano Gomez Luque

Tutored by Ingeborg Rocker

The Nomad, The Technologist

Wang Xiaowei

Tutored by Pierre Bélanger

MIT

Silk Pavilion

Markus Kayser, Jared Laucks,

Carlos David Gonzalez Uribe,

Jorge Duro-Royo

Tutored by Neri Oxman

Sketching in 3D

Woongki Sung

Tutored by Takehiko Nagakura

Post-Mortar Architecture

Rizal Muslimin

Tutored by Terry Knight

Pratt Institute

Fashioning Lines

Chung-Kuang Chao

Tutored by Philip Parker

The 6th Borough

Luana Faco Reis

Tutored by Ferda Kolatan, Carla Leitao,

David Ruy

Farming Extreme

Luke Cunnington

Tutored by Sulan Kolatan

Princeton

Transnational Brussels: A Ground for Action

Antonia Weiss

Tutored by Hayley Eber

From the Sound Up

Matthew L. Bertsch, Willem Boning

Tutored by Guy Nordenson

Mixed Reality Modelling

Ryan Luke Johns

Tutored by Jesse Reiser

RPI

Plexus

Ellen Wong

Tutored by Francis Bitonti

Normative Fluidity

Graham Billings

Tutored by Andrew Saunders

Affective Machines

Will Pyatt

Tutored by Casey Rehm

SCI-Arc

Carved in Stone

Ashley Holder

Tutored by Elena Manferdini

Low Fidelity

Erin Besler

Tutored by Andrew Atwood

Vulnerable Lines

Jinsa Yoon

Tutored by Elena Manferdini

UCLA

Mathilde

Danae Ledgerwood

Tutored by Jason Payne

Divergent Monoliths and the Mute Icon

Ho Man Yeung

Tutored by Georgina Huljich

Contorted Vault

Kimberly Daul

Tutored by Hadrian Predock, Georgina Huljich,

Jason Payne

U Michigan

Vertical Territories of Recursion

Justin Tingue, Andrew Wolking,

David M de Céspedes, Zuliang Guo

Tutored by Matias Del Campo, Adam Fure

Carbon Wound: Lightweight Composites

Megha Chandrasekhar, Christopher Mascari,

Brandon Vieth

Tutored by Glenn Wilcox, Wes Mcgee

Force-Active Architectural Assemblies

Tom Bessai

Tutored by Sean Ahlquist, Geoffrey Thun

U Penn

A Nightclub for Hong Kong

Andreas Koustapoulos, Hayley Wong

Tutored by Ali Rahim

Skirts

Tingwei Xu,Xie Zhang

Tutored by Ferda Kolatan

Plastic Lotus Robotic System

Mo Zheng, Eric Craig

Tutored by Winka Dubbeldam

USC

Alloplastic Architecture

Behnaz Farahi

Tutored by Alvin Huang, Neil Leach, Michael Fox

Poly-Directional Source fields

Yunyun Huang

Tutored by Biayna Bogosian

Controllable Reaction Diffusion System

Wendong Wang

Tutored by Alvin Huang

Yale

'Assembly One' Pavilion

David Bench, Zac Heaps, Jacqueline Ho,

Eric Zahn, Amy Mielke, John Taylor Bachman,

Nicholas Hunt, Seema Kairam, John Lacy,

Veer Nanavatty, Rob Bundy,

Raven Hardison, Matt Hettler

Tutored by Brennan Buck, Teoman Ayas,

Matthew Clark

Disheveled Geometries

Nicholas Kehegias, RJ Tripodi

Tutored by Mark Foster Gage

Wired Integrated Composite Knit

Sarah Gill, Jacqueline Ho

Tutored by Keller Easterling

Europe

AA

Swarm Works

Saman Dadgostar, Sofia Miranta Papageorgiou,

Akber A. Khan, Felipe Sepulveda Rojas

Tutored by Theodore Spyropoulos, Shajay Bhooshan,

Mostafa El Sayed, Manuel Jiménez García

Life Aquatech

Armando Bussey, Edward Roger,

Douglas Lückmann, Vichayuth Meenaphant,

Ana Margarita, Wang Zunig

Tutored by Robert Stuart-Smith, Tyson Hosmer

Reformation

Yitzhak B. Samun, Sobitha Ravichandran,

Anusha Tippa, Di Ding

Tutored by Patrik Schumacher,

Pierandrea Angius

Angewandte

Binary Cycles

Daniel Prost, Stefan Ritzer

Tutored by Hani Rashid, Brian De Luna,

Sophie Grell, Armin Hess,

Jorg Hugo, Sophie Luger,

Andrea Tenpenny, Reiner Zettl

Order + Complexity + Contradiction

Josip Bajcer.

Tutored by Zaha Hadid, Mario Gasser,

Christian Kronaus, Jens Mehlan,

Robert Neumayr, Patrik Schumacher, Hannes Traupmann.

Vertical Strip

Stephan Sobl

Tutored by Greg Lynn

Bartlett

Synthetic Constructability

Rodrigo Novelo Pastrana, Elina Christou,

an Dierckx, Nikola Papic,

Joanna Theodosiou,

Shahad Tamer, Amirreza Mirmotahari

Tutored by Alisa Andrasek, Daghan Cam,

Maj Plemenitas

Deep Texture

Stefan Bassing

Tutored by Daniel Widrig

Wireflies Project

Angelopoulou Dimitra, Diamanti Vasiliki,

Karantaki Meropi

Tutored by Jose Sanchez

CITA

Haven

Jonas Ersson

Tutored by Phil Ayres, Paul Nicholas,

Mette Ramsgard Thomsen,

Martin Tamke

Monument on Saltholmen

Mattias Lindskog

Tutored by Martin Tamke,

Mette Ramsgard Thomsen,

Paul Nicholas, Phil Ayres

Cascade of Luminescence

Sofia Adolfsson, Oskar Edström,

Jonas Ersson,

Margarita Isabel Huszár,

Jesper Wallgren,

Ida Katrine Friis Tinning

Tutored by Paul Nicholas, Martin Tamke,

Mette Ramsgard Thomsen, Phil Ayres

TU Delft

Agriflux

Anurag Bhattacharva

Tutored by Henriette Bier, Nimish Biloria,

Martin Sobota

A Food Hub for Paris

Manuel Zucchi

Tutored by Henriette Bier, Nimish Biloria,

Martin Sobota

Transferium Almere 2.0

Vladimir Ondejcik

Tutored by Nimish Biloria, Henriette Bier,

M.Sobota, L. Lignarolo

DIA

DUNElab

Asa Darmatriaji, Olga Kovrikova, Timothee Raison

Tutored by Christos Passas, Matias del Campo,

Alexander Kalachev

Load Reactive Morphogenesis

Sebastian Bialkowski

Tutored by Krassimir Krastev, Alexander Kalachev

Ferro Fluids

Arusyak Manvelyan, Kate Shelegon,

Alexander Amirov

Tutored by Neil Leach, Alexander Kalachev,

Karim Soliman

IAAC

Stigmergic Fibers

Jean Akanish, Alexander Dolan, Jin Shihui, Ali Yerdel

Tutored by Marco Poletto, Claudia Pasquero,

Alexandre Dubor

Interactive Spanning of the Besòs

Nuri Choi, Youssef Rashdan.

Tutored by Willy Muller, Hernan Diaz Alonso,

Peter Trummer, Maite Bravo

Anti-Gravity Object Modeling

Petr Novikov, Saša Joki

Tutored by Areti Markopoulou,

Luis Fraguada, Fabian Scheurer,

Mette Ramsgard Thomsen, Javier Peña

Innsbruck

Aggregated Lobbies

Michael Strobl, Ulrike Brandauer

Tutored by Peter Trummer, Ursula Frick,

Thomas Grabner

Aggregated Bundle Tubes

Pascal Leitgeb, Spirk Patrick Bayer

Tutored by Peter Trummer, Ursula Frick,

Thomas Grabner

Aggregated Floors

Simeon Brugger, Max Hohenfeller

Tutored by Peter Trummer, Ursula Frick,

Thomas Grabner

Paris Malaquais

Neuro Design

Cristina Osorio, David Ottlik, Rodolphe Bouquillard,

Jan Feichtinger, Estelle Glinel,

Clement Gosselin, Maxime Monin,

Mathieu Venot, Georgios Petros Lazaridis,

Louise Decroix, Bertrand Chapus,

Ludovic Haehnsen, Jules Salmon,

Adrien Taraki, Aljoscha Beiers,

Halna Boudet, Elisa Sepulveda Ruddoff,

Nicolas Dumas, Niels Barateig,

Nicolas Buffet,

Marie Lhuillier, Jessica Gerard.

Tutored by Pierre Cutellic

Regressive Resolutions 1-2

Charles Bouyssou, Lea Chang,

Nadja Gaudillere, Estelle Glinel,

Clement Gosselin, Corentin Heraud,

Antoine Le Dreff, Maxime Monin, Paul Poinet,

Mathieu Venot, Ioanna Zachariadou

Tutored by Pierre Cutellic, Sylvain Usai

ICD Stuttgart

Complex Tensile Surface Morphologies

Boyan Mihaylov, Viktoriya Nicolova,

Michael Pelzer, Christine Rosemann

Tutored by Sean Ahlquist, Achim Menges

Connecting Intelligence

Oliver David Krieg

Tutored by Achim Menges, Jan Knippers

Wood Laminate Morphologies

Bum Suk Ko

Tutored by Sean Ahlquist, Achim Menges

ETH Zurich

Brickolage

Katia Ageeva, Diana Alvarez,

Orestis Argyropoulos,

Stella Azariadi, Tianyi Chen,

Yun-Ying Chiu, Ivana Damjanovic,

García Pepo Martínez,

Melina Mezari, Bojana Miskeljin,

Evangelos Pantazis, Stanislava Predojevic,

Stylianos Psaltis, Meda Radovanovic,

Daniel Rohlek, Miro Roman,

Castro Mauricio Rodríguez,

Teemu Seppänen, Grete Soosalu

Tutored by Mathias Bernhard,

Manuel Kretzer,Tom Pawlofsky

High-Resolution

Jeanne Wellinger, David Jeanny,

Christian Suter, Li Bo,

Romain Frezza, Yann Bachofner,

Petrus Aejmelaeus-Lindström,Yushi Sasade

Tutored by Benjamin Dillenburger,

Michael Hansmeyer, Hua Hao

Resinance

Mark Baldwin, Jessica In,

Tihomir Janjusevic, Nan Jiang,

Joel Letkemann, Nicolas Miranda,

Irene Prieler, David Schildberger,

Demetris Shammas, Maria Smigielska,

Akihiro Tanigaito, Evi Xexaki,

Achilleas Xydis, Yuko Ishizu

Tutored by Manuel Kretzer,

Benjamin Dillenburger, Hironori Yoshida,

Weixin Huang, Lei Yu,

Tomasz Jaskiewicz, Mariana Popescu,

Andrei Pruteanu, Stefan Dulman

Australia

RMIT

FabPod

Daniel Davis, Alex Pena de Leon, Matthew Azzalin,

Aphiphong Chaitchavalit, Jihun Kang,

Thippanawat Sunantachaikool, Errol Xiberras,

Xuanqi Yang, Lu Ping, Tuyen Tran, Ciara McGrath,

Frank Mwamba, Robert Doe, Tom Hammond,

Heike Rahmann, Jeremy Ham

Tutored by Nick Williams, Brady Peters, John Cherrey,

Jane Burry, Mark Burry

Tangible Teamwork Table

Raul Kalvo, Davide Madeddu, Jakob Bak

Tutored by Flora Salim, Jane Burry, Sarah Pink,

Gerda Gemser, Kerry London

Composite Agency

May Yim Thaw, Stephanie Elena Lancuba,

Jedshamler Singudevan, Daniel Schulz,

Monique Mok, Anh Bui,

Arif Mustaqim, Hamnet Texgku,

Santiago Kojic, Wencheng Xi,

Nandana Nghaghskara Dermawan,

Errol Xiberras, James Pazzi,

Simon Glaister, Marc Gibson,

David Francis Smith,

Nil Corominas Faja

Tutored by Roland Snooks

China

HKU

Inclusive Occlusion

Elaine Choy

Tutored by David Erdman

Trigger (Arch)

Eric Wai Kin Lo

Tutored by Christian Lange

Climate Skin

Cai Hongkui

Tutored by Tom Verebes

HUST

Youth Hostel Design

Huang Junyan

Tutored by Tan Song

Department of Architecture Extension

Wang Zaoxuan

Tutored by Tan Song

HNK

Remodeling of Entrance Lobby

Wen Wei, Gao Xiaoyu, Xie Yihuan,

Cai Rubing, Li Zhixin

Tutored by Hu Biao, Du Yu, Wan Xinyu,

Johannes Elias

Large-Span Structure

Zhou Yu, Mao Lu, Dong Yina, Wen Wei,

Gao Xiaoyu,Zu Hanxing

Tutored by Hu Biao, Song Mingxing,

Lu Jiansong

Large-Span Space

Pei Zejun, Liao Shiwei, Yang Hengyu,

Mu Shiwen, Xie Yiheng, Lai Sichao

Tutored by Hu Biao, Song Mingxing,

Lu Jiansong

SCUT

Digital Scarpa

Lin Ruigu, Jin Yuan,

Zhang Mei, Yuan Xiaoyu

Tutored by Song Gang, Zhong Guanqiu

Generated High-Rise

Lin Rungu

Tutored by Song Gang

Thermo Surface

Dai Sen, Chen Yuchuan

Tutored by Song Gang, Su Aidi, Dai Xiaoben

Tianjin University

Revolve

An Yichen

Tutored by Xu Zhen

Holy Light

Yin Ting

Tutored by Xu Zhen

Canal Ark

Zeng Lingyue

Tutored by Xu Zhen

TJU

Micro Cities

Li Xuening, Xie Jie, Zhang Bowen,

Xiao Tong, Zhang Jia, Gong Yinjia

Tutored by Philip F. Yuan

Solar Decathlon Europe 2012

Yu Zhongqi, Qian Lie, Jia Dongfang,

Wu Xiaofan, Jin Dong, Luo Guofu,

Zhao Shijia, Cao Hanxiao,

Zhu Shengwei, Hu Yifeng,

Tang Wenwen, Zhou Bin,

Mu Di, Hou Jinming, Sun Hao,

Yang Guangyao, Lu Xinyi,

Bing Heiliang, Lei Yong, Xu Le,

Cao Ke, Liu Kang

Tutored by Tan Hongwei,

Qian Feng, Philip F. Yuan,

Wang Lingling, Li Qiang

Xixi Wetland Museum

Chen Jingcheng

Tutored by Philip F. Yuan

THU

Glue

Hao Tian, Huang Haiyang, Tan Sizhi

Tutored by Xu Weiguo, Lin Qiuda

DLA

Ao Liu, Chongxiao Xi

Tutored by Weiguo Xu,

Weixin Huang, Feng Xu

Chao-proliferate Polynomial Curves

Liang Yingya, Yu Haochang, Wu Tong

Tutored by Xu Weiguo, Huang Weixin

XAUAT

Topological Optimization

Su Kelun

Tutored by Jing Minfei, Wong Dong

Information and Technology Center

Wu Yanshan

Tutored by Jing Minfei

Haisen Fruit Headquarters

He Mincong

Tutored by Jing Minfei, Ye Fei, Wang Dong

作者简介 / BIOGRAPHIES

徐卫国 /Xu Weiguo

徐卫国教授执教于清华大学建筑学院，现为建筑系系主任；曾是美国麻省理工学院访问学者，2011—2012 年执教于美国南加州建筑学院 (SCI-Arc) 及南加州大学建筑学院 (USC)；并曾在日本留学获日本京都大学博士学位，工作于日本村野藤吾建筑事务所。他在任清华建筑学院教授的同时，建立 XWG 建筑工作室，从事建筑设计工作，并在多项设计竞赛中获奖。他是国内参数化非线性建筑设计的开拓者，发表论文 70 余篇，出版专著及编著 10 本。1999 年参加第 20 届 UIA 国际建筑师大会中国青年建筑师作品展，之后又在美国、法国、德国、意大利、俄罗斯、以色列、日本等国讲演或举办展览；他还是北京国际青年建筑师及学生作品展（2004、2006、2008、2010）的策展人；2012 年作为主要发起人，创立数字建筑设计专业委员会并被选为专委会主任；2013 年组织 DADA 系列活动。

Xu Weiguo is Professor and Chair of Architecture in the School of Architecture at Tsinghua University. He was a Visiting Scholar at MIT in 2007 and taught in SCI-Arc and USC in 2011-2012. He studied architecture at Tsinghua University, and then started teaching at the same institution before moving to Japan to work for Murano Mori Architects. He was awarded his doctorate from Kyoto University in Japan. On returning to China, he established his own architectural practice (XWG) in Beijing. He is the recipient of many awards, and his 70 works have been published in many journals. He is the author of 10 books including The Way of Architectural Design Thinking (China Architecture and Building Press, 2001), Architecture/Non-Architecture (China Architecture and Building Press, 2006), Studio Works of Tsinghua Students (Tsinghua University Press, 2006), and co-author of Fast Forward >> (MAP Books, 2004), Emerging Talents, Emerging Technologies (2 Vols., China Architecture and Building Press, 2006), (Im)material Processes: New Digital Techniques for Architecture (2 Vols., China Architecture and Building Press, 2008), Machinic Processes (2 Vols., China Architecture and Building Press, 2010). Xu Weiguo was included in the Exhibition of Young Chinese Architects at UIA Congress 20th in 1999, and was selected as one of the architects to represent China in the A1 Pavilion at ABB 2004. He was one of the curators of Architecture Biennial Beijing 2004, 2006 2008 and 2010. He has had lectures and exhibitions in different countries including USA, France, Germany, Italy, Israel, Russia, and Japan. As one of main initiators, he established the Digital Architecture Design Association and was elected Director of DADA in 2012. He organized the DADA series events in 2013.

尼尔·林奇 /Neil Leach

尼尔·林奇是一位建筑师兼理论家，他目前是南加州大学建筑系的教授，同时他还是美国航天局高级创新专员。他曾经执教于包括南加州建筑学院、伦敦建筑师联盟建筑学院、康奈尔大学、哥伦比亚大学建筑学院、西班牙加泰罗尼亚高级建筑研究所、德国德绍建筑学院、巴斯大学、布莱顿大学和诺丁汉大学，其著作包括《空间政治》（劳特利奇出版社即将出版）、《伪装》（麻省理工学院2006年出版）、《忘掉海德格尔》（派迪亚出版社2006年出版）、《中国》（香港麦普奥菲斯出版社2004年出版）、《千年文化》（伊利普西思出版社1999年出版）和《建筑麻醉学》（麻省理工学院2006年出版）；曾与他人合著《玛思潘滋》（建筑基金出版社2000年出版）；曾编辑《数字城市》（怀利出版社即将出版）、《为数字世界而设计》（怀利出版社2002年出版）、《空间的象形文字》（劳特利奇出版社2002年出版）、《建筑与革命》（劳特利奇出版社1999年出版）和《建筑的反思》（劳特利奇出版社1997年出版）；与他人共同编辑《集群智能：多代理系统建筑》（同济大学出版社2012年出版）；《探访中国数字建筑设计工作营》（同济大学出版社2013年出版）、《建筑数字化编程》（同济大学出版社2012年出版）、《建筑数字化建造》（同济大学出版社2012年出版）、《数字建构：青年建筑师/学生设计作品》（两册，中国建筑工业出版社2010年出版）、《涌现：青年建筑师/学生设计作品》（两册，中国建筑工业出版社2006年出版）、《快进·热点·智囊组》（香港麦普奥菲斯出版社2004年出版）、《数字建构》（怀利出版社2004年出版）；还是阿尔伯蒂的《建筑艺术十书》（麻省理工学院1988年出版）的译者之一。他与徐卫国共同策展了一系列的北京建筑双年展：快进 >>（2004）、涌现（2006）、数字建构（2008）、数字现实（2010），同时与 Roland Snooks 共同策展了2010年上海画廊“集群智能：多代理系统建筑”，与袁烽共同在同济大学策展了数字未来展览（2011）以及互动上海展览（2013）。他目前正在参与由美国航天局赞助的研究项目。此项目着眼于开发可用在月亮及火星上三维打印结构的智能机器人。

Neil Leach is an architect and theorist. He currently teaches at the University of Southern California, and is a NASA Innovative Advanced Concepts fellow. He has also taught at SCI-Arc, Architectural Association, Cornell University, Columbia GSAPP, Dessau Institute of Architecture, Royal Danish School of Fine Arts, IaaC, ESARQ, University of Bath, University of Brighton and University of Nottingham. He is the author of The Politics of Space (Routledge, forthcoming), Camouflage (MIT, 2006), Forget Heidegger (Paideia, 2006), China (Map Office, 2004), Millennium Culture (Ellipsis, 1999) and The Anaesthetics of Architecture (MIT, 1999); co-author of Mars Pants: Covert Histories, Temporal Distortions, Animated Lives (Architecture Foundation, 2000); editor of Digital Cities (Wiley, 2009), Designing for a Digital World (Wiley, 2002), The Hieroglyphics of Space: Reading and Experiencing the Modern Metropolis (Routledge, 2002), Architecture and Revolution: Contemporary Perspectives on Central and Eastern Europe (Routledge, 1999), and Rethinking Architecture: A Reader in Cultural Theory (Routledge, 1997); co-editor of Swarm Intelligence: Architectures of Multi-Agent Systems, (Tongji UP, forthcoming); Digital Workshop in China (Tongji UP, 2013) Scripting the Future (Tongji UP, 2012), Fabricating the Future (Tongji UP, 2012); Machinic Processes, 2 Vols. (CABP, 2010), (Im)material Processes: New Digital Techniques for Architecture, 2 Vols., (CABP, 2008), Emerging Talents, Emerging Technologies, 2 Vols., (CABP, 2006), Fast Forward>>, Hot Spots, Brain Cells (Map Office, 2004) and Digital Tectonics (Wiley, 2004); and co-translator of Leon Battista Alberti, On the Art of Building in Ten Books (MIT Press, 1988). He has also been co-curator (with Xu Weiguo) of a series of exhibitions at the Architecture Biennial Beijing: Fast Forward>> (2004), Emerging Talents, Emerging Technologies (2006), (Im)material Processes: New Digital Techniques for Architecture (2008), and Machinic Processes (2010). He also co-curated (with Roland Snooks) Swarm Intelligence: Architectures of Multi-Agent Systems in SH Gallery, Shanghai (2010), and (with Philip Yuan) DigitalFUTURE (2011) and Interactive Shanghai (2013) in Tongji University, Shanghai. He is currently working on a research project sponsored by NASA to develop a robotic fabrication technology to print structures on the Moon and Mars.

图书在版编目（CIP）数据

设计智能 高级计算性建筑生形研究 学生建筑设计作品 / 徐卫国，(英)尼尔·林奇编. —北京：中国建筑工业出版社，2013. 10
ISBN 978-7-112-15842-3

Ⅰ. ①设… Ⅱ. ①徐…②尼… Ⅲ. ①智能化建筑—建筑设计—作品集—世界—现代 Ⅳ. ① TU243

中国版本图书馆 CIP 数据核字（2013）第 215612 号

责任编辑：张 建 施佳明
责任校对：姜小莲 王雪竹

设计智能
高级计算性建筑生形研究
学生建筑设计作品
徐卫国 尼尔·林奇（英）编
*
中国建筑工业出版社 出版、发行（北京西郊百万庄）
各地新华书店、建筑书店经销
北京画中画印刷有限公司印刷
*
开本：889×1194 毫米 1/20 印张：11 1/5 字数：336 千字
2013 年 10 月第一版 2013 年 10 月第一次印刷
定价：98.00 元
ISBN 978-7-112-15842-3
(24616)